SHELLS
OF THE WORLD

SHELLS OF THE WORLD
A NATURAL HISTORY

M. G. Harasewych

PRINCETON UNIVERSITY PRESS
PRINCETON AND OXFORD

Published in 2024 by Princeton University Press
41 William Street, Princeton, New Jersey 08540
99 Banbury Road, Oxford OX2 6JX
press.princeton.edu

Copyright © 2024 by Quarto Publishing plc

Conceived, designed, and produced by
The Bright Press
an imprint of The Quarto Group
1 Triptych Place, London, SE1 9SH, United Kingdom
www.Quarto.com

All rights reserved. No part of this book may be reproduced or transmitted in any form or by any means, electronic or mechanical, including photocopying, recording, or by any information storage-and-retrieval system, without written permission from the copyright holder. Requests for permission to reproduce material from this work should be sent to permissions@press.princeton.edu

Library of Congress Control Number: 2023939566
ISBN: 978-0-691-24827-1
Ebook ISBN: 978-0-691-24825-7
British Library Cataloging-in-Publication Data is available

Publisher **James Evans**
Editorial Director **Isheeta Mustafi**
Art Director **James Lawrence**
Managing Editor **Jacqui Sayers**
Senior Editors **Joanna Bentley, Caroline Elliker**
Project Editor **Ruth Patrick**
Design **Kevin Knight**
Picture Research **Katie Greenwood**
Illustrations **John Woodcock**

Cover photos: Front (clockwise from top left): 1-5, 7, 9, 11, 12, 14, 15: M. G. Harasewych; 6, 13: H. Zell, CC BY-SA; 8: Muséum de Toulouse/Didier Descouens, CC BY-SA; 10, 14: © Guido & Philippe Poppe—www.conchology.be. Back cover and spine: M. G. Harasewych.

Printed in Malaysia

10 9 8 7 6 5 4 3 2 1

6 Introduction

CONTENTS

16 **The Shelled Classes**

18 Class Polyplacophora

34 Class Monoplacophora

36 Class Bivalvia

100 Class Scaphopoda

108 Class Gastropoda

210 Class Cephalopoda

224 Appendices
226 Glossary
231 Further Reading
233 Index
239 Picture Credits
240 Acknowledgments

RIGHT | *Conus geographus*, a carnivorous snail that injects venom into its prey through a hollow, harpoon-like tooth. Its venom is fatal to humans within minutes.

INTRODUCTION

Shells (the external skeletons of mollusks) have intrigued and inspired humans and their ancestors from the Paleolithic to the present. Through the ages, shells have been used as tools, jewelry, decorations, currency, and religious symbols, and have inspired many forms of art, including architecture. Many of the mollusks that produced them have served as food, contributed to human illnesses (generally by being hosts to toxic microorganisms or intermediate hosts of human parasites), and are used in traditional medicine as treatments for a variety of ailments in many parts of the world.

SHELL FORMATION

A variety of organisms, among them foraminifera, corals, arthropods, and turtles, produce calcareous external coverings (mostly or partly composed of calcium carbonate) to protect the animal from the environment. However, the terms "shell" or "seashell" are limited to the external skeletons of mollusks. Arthropods produce an exoskeleton that includes an outer cuticle, which is periodically shed at certain stages in their life cycles (this process is known as ecdysis) or as the animal grows (molting), while the bones of vertebrate skeletons contain living cells and are resorbed and regenerated during the life of the animal. In contrast, the shells of mollusks remain a part of the animal throughout its life. And while all seashells are produced by mollusks, not all mollusks living today produce shells. Several lineages that evolved from shelled ancestors now have reduced internal shells or no longer produce shells. Most notable are terrestrial slugs, nudibranchs, and most cephalopods.

A mollusk shell is first formed during the larval stage and continues to grow by addition of shell material along its outer edge, in the process archiving the environment of the mollusk at the time of deposition. Much like tree rings, these layers can be studied to determine the life span of the individual mollusk, which may range from less than a year to several centuries. The bivalve *Arctica islandica* is the longest living individual animal known; one specimen having lived to the age of 507 years before it was collected and analyzed (see page 92).

Trace elements and isotopes from the surrounding environment are incorporated into the deposited shell layers at the time that each shell increment is secreted. They can be analyzed to determine the environmental conditions at the time the mollusk lived as well as the age of the shell, which, if fossilized, can range in the hundreds of millions of years.

The shell is secreted outside the animal's tissues by the outer surface of the mantle, an organ that is present in all mollusks. One section secretes a layer of protein (periostracum), while other cells secrete a fluid rich in calcium carbonate into the narrow space between the periostracum and the mantle. This fluid crystallizes onto the inner surface of the periostracum to produce the mineralized shell.

Rates of shell formation vary among mollusks. Some grow in small, regular increments; others form periodically in large sections. Growth tends to be rapid until the mollusk reaches adulthood. In some mollusks, growth continues at a much slower rate. In others, such as cowries and spider conchs, shell shape changes significantly in adulthood and growth is limited to small increases in thickness.

BELOW | A section through the shell of *Nautilus pompilius* showing that the shell grows by addition of material to its outer edge in the form of a logarithmic spiral.

THE SUCCESS OF MOLLUSKS

Mollusks—with and without shells—are among the most successful organisms and have diversified to occupy nearly all habitats. They originally evolved in the oceans of the world and now occur in all marine habitats from the splash zones above high-tide lines to the deepest ocean trenches, including hydrothermal vents, and from the tropics to both polar regions. Several lineages of gastropods and bivalves have independently adapted to freshwater habitats, from estuaries to rivers, streams, lakes, and hot springs. Other gastropod lineages have evolved the ability to breath air and live on land, ranging from nearshore environments to grasslands, trees, mountains, and deserts. Some squid species are even capable of briefly flying, much like flying fish, by swimming rapidly near the surface and then leaving the water to become airborne for tens of feet to escape predators.

Several mollusks are among the largest known animals. Some gastropods and bivalves approach 40 in (1 m) in length. Giant clams can attain a weight of 440 lbs (200 kg) while giant and colossal squid can exceed 50 ft (15 m) in length and 1,100 lbs (500 kg) in weight. Nevertheless, most mollusks are small animals. A survey of the mollusks collected in New Caledonia, for example, revealed that the majority of the nearly 3,000 species sampled had an adult shell size that was less than $5/8$ in (17 mm). Some species never grow larger than $1/32$ in (1 mm) in length.

Mollusks are extremely diverse in their habitats. The majority of aquatic species are benthic, living upon or burrowing within the ocean bottom. Many of these species have larvae with a planktonic stage (floating in the open ocean) that later metamorphose, or transform, and settle to the ocean bottom. Some mollusks, including most cephalopods (such as all Nautilidae, Spirulidae, Loliginidae, and some octopods like Argonautidae) and some gastropods (such as Carinariidae and Cavoliniidae), are pelagic, spending their entire lives in the open ocean as free-swimming animals. Among the benthic animals, some occur in high densities and form reefs (for example, oyster reefs), which can serve as a habitat for many other animals. Others occur more sparsely, their densities often determined by their diet and food availability.

LEFT | *Tridacna gigas* releasing spawn into the water along the Great Barrier Reef, Queensland, Australia.

THREATS TO MOLLUSKS

Despite their diversity, many shelled mollusks have been and continue to be negatively affected by a variety of human activities, ranging from overfishing and introduction of invasive species that compete with native species to habitat destruction (for example, deforestation, dam construction, and oil spills) and increasing pollution of the land, sea, and air. The most immediate and greatest risk is for shelled mollusks that are limited to small or endemic populations, most notably terrestrial snails known only from small islands as well as freshwater snails and bivalves that are endemic to single springs, streams, or drainages.

The International Union for Conservation of Nature (IUCN) maintains a Red List of Threatened Species, which documents the extinction risk of portions of the world's biota (plant and animal life within a particular area). It is strongly biased toward mammals and birds, but includes data on 9,862 species of mollusks, listing 308 species as extinct, 19 as extinct in the wild, and 744 as critically endangered. Human activities affect ocean-dwelling species but have generally been limited to areas that are smaller than the range of impacted species, which survive outside those areas, and can, over time, repopulate them. However, even mollusks that are endemic to hydrothermal vents—small hot springs along the ocean floor at depths of several thousand feet—are included in the Red List. Of 184 such endemic species that have been assessed, 39 are listed as critically endangered and another 75 endangered or vulnerable due to deep-sea mining for rare minerals.

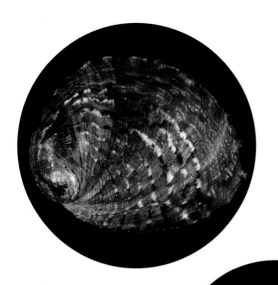

ABOVE | The Northern Abalone *Haliotis kamtschatkana* has been listed as an endangered species since 2006, according to the IUCN Red List.

RIGHT | Free-swimming planktonic larvae of *Ostrea edulis*. After metamophosis, they will settle and become permanently attached to a hard substrate.

Modes of feeding also vary. Some feed on detritus, others are filter feeders; some are herbivores scraping algae with their radula (a flexible ribbon with multiple, rasp-like teeth). Many lineages of carnivores have evolved specialized feeding mechanisms. Some (such as Naticidae and Muricidae) are capable of penetrating the shells of their prey using a combination of chemical and mechanical means; others inject prey with toxins that rapidly paralyze them. Still others are parasitic on other mollusks or echinoderms (marine invertebrates that include starfish, sea urchins, and sea cucumbers). Giant clams host symbiotic algae within their tissues that contribute to their nutrition, while bivalves that occur at hydrothermal vents in the deep sea rely on symbiotic chemosynthetic bacteria growing within their gills for their nutrition.

THE ORIGIN AND EVOLUTION OF MOLLUSKS

All mollusks, living and extinct, are descended from a common ancestor. The phylum Mollusca is among the oldest groups of organisms on Earth, with many of the major shelled lineages already well-represented in the early Cambrian (539 mya) fossil record. Several taxa that either lacked a shell or had a flexible or weakly mineralized shell were present in the pre-Cambrian (555 mya) fauna and some have been questionably assigned to the phylum Mollusca.

MOLLUSK DIVERSITY

Mollusks have survived five major mass extinctions over the past 450 million years, each of which caused the extinction of over 75 percent of all animals and plants living at the time. Although multiple molluscan lineages perished during these extinctions (among them Rostroconchia, Bellerophontoidea, and Ammonoidea), those that survived proliferated to become the most diverse group of animals in the oceans, and the second-most diverse phylum living today, second only to Arthropoda, which includes insects.

Estimates of molluscan diversity are in the range of 80,000 to over 100,000 presently known living species, with nearly as many extinct fossil species. Other estimates suggest that fewer than half of living species of mollusks have thus far

BELOW | Drawing of an animal of the Ammonoid order Goniatida, which evolved during the Middle Devonian (390 mya) and became extinct during the End Permian Extinction (251.9 mya).

RIGHT | Ammonite fossil of Jurassic age exposed on Monmouth Beach, West of Lyme Regis, Dorset, England.

been discovered and documented. This extraordinary diversity of the phylum Mollusca is currently partitioned into eight classes living today, each representing a separate lineage that traces its ancestry, directly or indirectly, to the ancestral mollusk.

The evolutionary relationships among these lineages are the subject of ongoing research and remain in flux, undergoing frequent revisions and refinements as more data, increasingly based on DNA sequences, are accumulated. Phylogenomic studies indicate an early division within the phylum into two lineages: the Aculifera and the Conchifera. The Aculifera, which include the two shell-less lineages Caudofoveata and Solenogastres (currently regarded as separate classes, but previously considered to be subclasses within the class Aplacophora) as well as the Polyplacophora, which have a dorsal shell composed of eight interlocking valves, are thought to have originated at least 540 mya during the Ediacaran, while its sister taxon, the Conchifera, which includes the five remaining molluscan lineages that are descended from a single-shelled ancestor, appeared and diversified during the early Cambrian. The last common ancestor of all mollusks, the molluscan prototype, is thought to have lived during the Ediacaran (635–541 mya).

THE ANCESTRAL MOLLUSK

The hypothetical ancestral mollusk is deduced to have been a small (< 3 mm) animal that was bilaterally symmetrical along an anteroposterior axis, with an anterior head, a mouth containing a radula, a ventral foot, and a dorsal secretory mantle that was expanded posteriorly to form a mantle cavity that contained paired gills, osphradia, anus, as well as openings of the nephrida and genital organs. These animals had a dorsal chitinous cuticle in which were embedded spicules or scales. These ancestral mollusks are thought to have lived in marine habitats at sublittoral depths. They were likely microcarnivores. Reproduction was by external fertilization of large, yolky eggs that developed into planktonic larvae before settling to a benthic habitat.

ABOUT THIS BOOK

ABOVE | The common (European) cuttlefish (*Sepia officinalis*) is generally found in the eastern North Atlantic and Mediterranean Sea. It is a cephalopod, related to squid and octopus.

BELOW | Chart showing the ocean depths referred to in this book.

This book focuses on the shelled mollusks that inhabit the oceans of the world. Neither shell-less mollusks nor freshwater or terrestrial mollusks are featured in The Shelled Classes (see pages 16–223). A short introduction is provided on the shell-less classes on pages 14–15.

Of the more than 700 families within Mollusca apportioned among the six shelled classes, 93 examples have been selected to provide a broad overview of the immense variability in morphology, habitat, and behavior within the phylum Mollusca. These are arranged according to our current understanding of the phylogenetic relationships of the classes and the families within the classes. Ongoing research to document molluscan biodiversity results in the discovery of dozens to hundreds of new species annually, while multiple studies of the relationships result in frequent additions and rearrangements

OCEAN ZONES

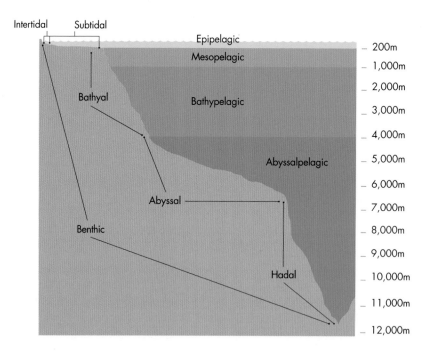

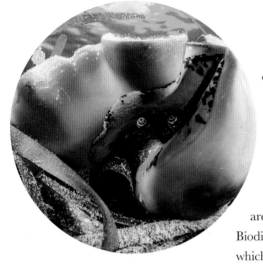

of genera and families within Mollusca. Websites such as the World Register of Marine Species (www.marinespecies.org) and Molluscabase (www.molluscabase.org) provide updated accounts of our current understanding of molluscan diversity and interrelationships of the various higher taxa.

The maps showing the distribution of each family are based on records documented by the Global Biodiversity Information Facility (GBIF, www.gbif.org), which is also continuously updated.

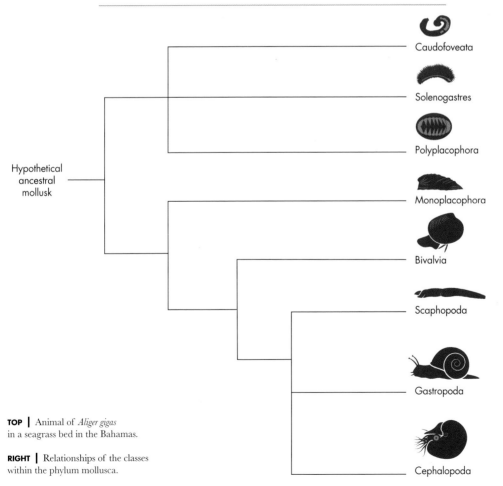

TOP | Animal of *Aliger gigas* in a seagrass bed in the Bahamas.

RIGHT | Relationships of the classes within the phylum mollusca.

13

THE SHELL-LESS CLASSES

CLASS CAUDOFOVEATA

Neither common nor diverse, Caudofoveata are known from a single order containing 3 families, 14 genera, and about 150 species. The Caudofoveata are considered to include the most primitive mollusks living today. They exhibit most of the features attributed to the hypothetical ancestral mollusk.

Caudofoveata are worm-shaped and covered by a cuticle composed of protein, in which are embedded monocrystalline spicules. They have a cuticular oral shield, a simple bipartite radula, lack a ventral furrow or foot, and have one pair of gills (ctenidia) in the posterior mantle cavity. Caudofoveata have separate sexes and are broadcast spawners, releasing their sperm and eggs into open water. Eggs have a large yolk that provides nutrition to the developing trochophore (type of free-swimming, spherical, or pear-shaped marine larvae with cilia) before they settle to the ocean bottom.

ABOVE | Close-up of mouth of Caudofoveata *Falcidens* sp, from New Zealand, taken using a scanning electron microscope.

BELOW | *Chaetoderma argenteum*. Observed at a depth of 597 ft (182 m), soft bottom, off Tacoma, Washington.

Most species are less than an inch in length, although some grow to 5–6 in (125–150 mm). The majority of species have been reported from the northern hemisphere, where they burrow in the upper inch or so of soft sediments at depths ranging from 98–>23,000 ft (30–>7,000 m) and feed on organic detritus or on small organisms, primarily foraminiferans and small polychaetes (marine worms). This class has thus far not been recognized in the fossil record.

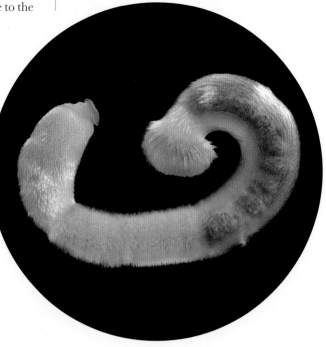

CLASS SOLENOGASTRES

Species of Solenogastres are also worm-shaped but are laterally compressed, with the ventral foot situated in a groove. The mantle produces a cuticle that has aragonitic spicules and scales. Solenogastres have an anteroventral mouth. A radula with thirteen teeth per row is present in most, but not all species. They lack gills or kidneys and are protandrous hermaphrodites. Specimens are born male, becoming female as they grow. Fertilization is internal. Fertilized eggs may be released or brooded in the mantle cavity.

Solenogastres currently contain 24 families, 75 genera, and 320 species. They occur in all oceans, although most species are known from the southern hemisphere. They inhabit shallow to abyssal (14,760 ft / 4,500 m) depths. Many are associated with hydroids, anemones, and soft corals, on which they feed. A putative Solenogastres has been reported in Silurian (443.8–419.2 mya) deposits.

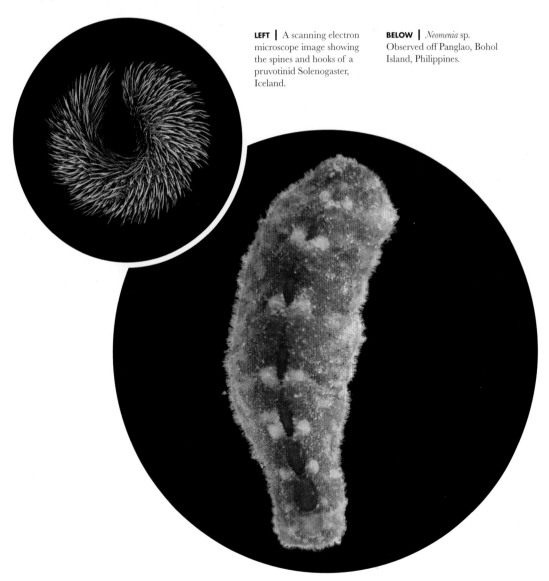

LEFT | A scanning electron microscope image showing the spines and hooks of a pruvotinid Solenogaster, Iceland.

BELOW | *Neomenia* sp. Observed off Panglao, Bohol Island, Philippines.

OPPOSITE | Clockwise from top left: *Tridacna noae*, *Melo georginae*, *Meiocardia moltkiana*, and *Cypraecassis rufa*.

THE SHELLED CLASSES

Six of the eight classes within the phylum Mollusca have calcified external shells. Of these, the class Polyplacophora (see page 18), with a shell composed of eight interlocking plates, is currently regarded as being most closely related to the shell-less classes Caudofoveata (see page 14) and Solenogastres (see page 15) and included in the subphylum Aculifera. The remaining five classes, Monoplacophora (see page 34), Bivalvia (see page 36), Scaphopoda (see page 100), Gastropoda (see page 108), and Cephalopoda (see page 210), are classified as members of the subphylum Conchifera, and share a common ancestor with a single shelled ancestor resembling members of Monoplacophora.

This chapter features a small subset of the more than 700 families that span the range of shelled mollusks inhabiting the world's oceans, with representation roughly proportional to the diversity among the classes. Some include common and wide-ranging genera and species that will be familiar to the casual observer. Others feature animals that provide insights into the broad ranges of habitats, behaviors, diets, and reproductive strategies that have and continue to evolve within the most diverse phylum in the oceans of the world.

OPPOSITE | *Mopalia ferreirai* crawling on rock in Queen Charlotte Strait, British Columbia, Canada.

POLYPLACOPHORA

Polyplacophorans, also known as chitons or coat-of-mail shells, are elongate, bilaterally symmetrical, dorsoventrally compressed animals. Their shells are secreted as eight separate, parallel, interdigitating valves: a head valve, six intermediate valves, and a tail valve, all of which can move relative to each other. These valves are held together by muscles and an elliptical, cuticularized girdle, allowing the animal to cling and conform to uneven surfaces and to roll into a ball if dislodged. Each shell plate is composed of four layers—the outermost layer is the periostracum, and is composed of protein. The remaining three layers—the tegmentum, the articulamentum, and the hypostracum—are all composed of crystalline calcium carbonate. Light-sensing organs (aesthetes) that are unique to chitons pass through narrow pores in the shell plates. The girdle may have scales, spines, spicules, or chitinous hairs.

Polyplacophorans have a large foot flanked by a mantle groove that contains up to forty pairs of gills (ctenidia). Gonopores (reproductive organs) and kidney ducts also open into the mantle groove. The anterior head bears a mouth but lacks tentacles or eyes. The anus is posterior. The radula, specialized for scraping hard substrates, contains 13–17

ABOVE | Shell of *Chiton tuberculatus*. The girdle is not present in this photo.

teeth per row, with 25–150 rows. Two teeth in each row are capped with magnetite (iron oxide) to make them harder and resist wear.

Sexes are separate in most chitons, although hermaphroditism has been reported in some species. Eggs are rich in yolk, and generally shed into the water or are deposited in clumps or gelatinous strings. Some species brood their eggs within the pallial groove.

Chitons are exclusively marine and live on hard substrates from intertidal to abyssal depths. Most live in shallow water on rocky shores and graze on algae, diatoms, foraminifera, and sponges growing on the rocks, but several taxa in three separate lineages are predators that capture and consume small crustaceans and worms. Some species exhibit homing behavior, returning to the same spot after feeding excursions. Deep-sea species feed on decaying sunken wood or biofilms that grow on it.

Most polyplacophoran species range in size from a fraction of an inch to about 4 in (100 mm), but some exceed 12 in (300 mm) in length. Living Polyplacophora are divided into 3 orders, 19 families, and approximately 950 species. Polyplacophora are represented in the fossil record since the Late Cambrian (490 mya).

SHELL FEATURES AND ANATOMY OF A TYPICAL POLYPLACOPHORAN

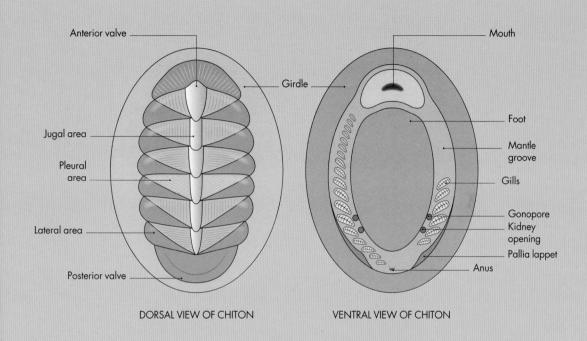

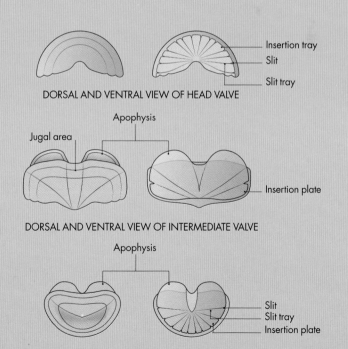

LEPIDOPLEURIDA—LEPTOCHITONIDAE
SLENDER CHITONS

The family Leptochitonidae, a representative of Lepidopleurida, the most basal order of living Polyplacophorans, occurs in all oceans of the world. It is comprised of mostly small species that are less than 1 in (25 mm) in length. Shell valves lack insertion plates and sutural lamellae are small. The tegmentum tends to be colorless and pustulose, while the girdle may be covered with pustules or overlapping scales. Gills are situated in the posterior portion of the mantle groove.

Most species inhabit deep water (to 26,250 ft/ 8,000 m), where they feed on bottom detritus and small organisms. Some deep-sea species feed on decaying wood and the biofilm growing on it.

Earliest records of this family, which represent an extinct lineage, date to the Ordovician (450 mya) deposits in Europe and North America. The majority of the known species are in the genus *Leptochiton*.

LEFT | *Leptochiton asellus* lives on the underside of rocks at depths to 820 ft (250 m) and is widespread throughout the North Atlantic Ocean.

OPPOSITE | *Leptochiton rugatus* on intertidal rock, Strathcona, British Columbia, Canada.

DISTRIBUTION
Global, from the tropics to polar seas. Some species are subtidal, but many occur at bathyal to abyssal depths.

DIVERSITY
Family includes approximately 150 living species assigned to 6 genera within 1 subfamily.

HABITAT
Shallow-water species live on or under rocks and hard substrates. Deep-water species may live on sunken wood or sponges.

SIZE
Species range in size from ¼–1¼ in (7–30 mm).

DIET
Grazers, feeding on detritus, small organisms, sunken wood, and the biofilms growing on it.

REPRODUCTION
Sexes are separate, and fertilization is external, with eggs and sperm shed into the water. The eggs are surrounded by a simple and smooth jelly-like hull. There is an extended pelagic, free-swimming larval stage prior to metamorphosis and settlement to the ocean bottom.

CALLOCHITONIDA—CALLOCHITONIDAE
CALLOCHITONS

The order Callochitonida, which contains a single family Callochitonidae, is the second-most basal lineage of living chitons. It diverged from the order Chitonida, which includes all remaining living chitons during the early Permian.

BELOW | *Callochiton septemvalvis* from Catalonia, Spain.

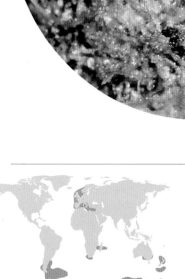

Shells of Callochitonidae are small to moderately large and ovate to elongate in outline. The tegmentum may be smooth, finely granulose, or with longitudinal ribs. There are multiple well-developed pigmented aesthetes. The insertion plates on the head and tail valves may have from ten to twenty-four slits, the median valves from one to four slits. The girdle is thick and covered with closely packed minute scales, spicules, or scattered longer needles or bristles. Gills span the length of the mantle groove.

Callochitonidae live on hard substrates in shallow waters to depths of over 1,640 ft (500 m). Some species preferentially feed on coralline algae. Sexes are separate. Fertilization is external, with eggs covered with a jelly-like layer. Eggs hatch into lecithotrophic planktonic larvae without a veliger stage before metamorphosis.

DISTRIBUTION
Occurs in European seas, subtropical and tropical regions of the Indo-Pacific, and along the Southern Ocean.

DIVERSITY
Family includes 51 living species assigned to 5 genera within 1 subfamily.

HABITAT
On or under rocks and hard substrates at intertidal to bathyal depths up to 1,640 ft (500 m).

SIZE
Species range in size from $^5/_{16}$–4½ in (8–114 mm).

ABOVE | *Eudoxochiton nobilis* from the Bay of Plenty, New Zealand.

RIGHT | *Callochiton sulcatus* from the Galapagos Islands.

DIET
Grazers, feeding on rocks covered with algae, diatoms, and detritus.

REPRODUCTION
Sexes are separate and fertilization is external. Larvae have a free-swimming stage before settling to the ocean bottom.

CHITONIDA—CHITINOIDEA—CHITONIDAE
SCALED CHITONS

Members of the family Chitonidae can reach 8 in (200 mm) in length. All the valves have well-developed insertion plates with slits separating comb-like teeth along the valve margins. Tegmentum sculpture ranges from finely to coarsely granulose and may be strongly grooved. Sutural lamellae are well-developed, and the girdle is covered with closely packed scales. The gills span the length of the pallial grooves and the major teeth of the radula have a single, broad cusp.

Many species occur in high densities on intertidal and subtidal rocky surfaces. They tend to feed nocturnally, grazing on algae, sponges, and other encrusting organisms. Several species exhibit homing behavior, returning to their resting place after nocturnal feeding excursions. Life span ranges from two to twelve years in some species.

Earliest records of the family Chitonidae date to the Carboniferous (300 mya).

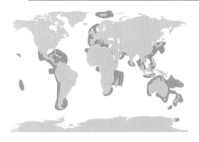

DISTRIBUTION
Global, with most species inhabiting temperate and tropical latitudes, but some occur in polar regions.

DIVERSITY
Family includes approximately 200 living species assigned to 13 genera within 3 subfamilies.

HABITAT
On or under rocks and other hard substrates, from intertidal to subtidal depths.

SIZE
Most species range in size from 1–3 in (25–75 mm) but several species may reach 6–8 in (150–200 mm).

ABOVE | *Rhyssoplax olivacea*, a species that ranges throughout the Mediterranean and along the Atlantic coasts of Portugal and Morocco, on submerged rocks.

OPPOSITE | *Chiton tuberculatus*, a large chiton from the Caribbean.

DIET
Grazing omnivores, feeding on various algae, including coralline algae, but also small encrusting organisms.

REPRODUCTION
Sexes are separate and most species are free spawning, but some (e.g. *Radsia*) brood their young within the pallial groove until they are crawling juveniles. Eggs have hulls with spine-like projections.

ABOVE | *Acanthopleura spinosa*, a common species in the central Indo-Pacific, clinging to a rock at low tide.

CHITONIDA—CRYPTOPLACOIDEA—ACANTHOCHITONIDAE
PRICKLY CHITONS

The family Acanthochitonidae is represented in the living fauna by 10 genera and nearly 200 species within the subfamily Acanthochitoninae, and by a single genus with one species in the subfamily Cryptochitoninae. Members of the subfamily Acanthochitoninae are diverse. They live on hard substrates such as rocks, rubble, reefs, shells, and mangroves, mostly in shallow water, although some genera reach depths of 1,640 ft (500 m).

The shell valves are articulated, the tegmentum may be reduced, and the girdle is fleshy with spicules and prominent tufts at the sutures of the valves. The gills are confined to the posterior portion of the

DISTRIBUTION
Global, from the tropics to polar regions, from intertidal to bathyal depths.

DIVERSITY
Family includes approximately 200 living species assigned to 11 genera within 2 subfamilies.

HABITAT
On or under rocks and hard substrates.

SIZE
Most species range in size from 1–4 in (25–100 mm) but one species reaches 16 in (400 mm).

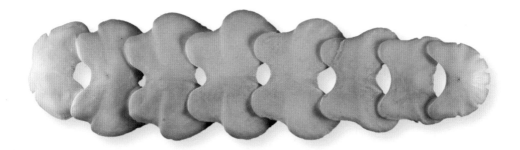

ABOVE | The exposed internal shell of *Cryptochiton stelleri*, composed of eight plates that are loosely interconnected.

BELOW | *Cryptochiton stelleri* feeding on seaweed exposed at low tide off Newport, Oregon.

mantle grooves and the major teeth of the radula have three cusps. Sexes are separate and gametes are shed into the water. Eggs have elaborately sculptured hulls with large projections with thick bases.

Cryptochiton stelleri, the only species in the subfamily Cryptochitoninae, is the largest species of chiton. It may reach a length of up to 16 in (400 mm) and a weight of 2 lbs (900 g). It is the only chiton in which all the shell plates are internal, having been overgrown by the girdle. The valves lack a tegmentum. It lives in shallow water on rocky shores in subarctic regions of the North Pacific from Japan to the western coast of North America. This species is known to live for up to twenty-five years.

OPPOSITE | *Acanthochitona fascicularis*, a species with prominent spicules along its girdle, on a rocky coast in Brittany, France.

DIET
Grazers, feeding on algae. Some species feed on sponges; others capture small crustaceans and worms by trapping them under an anteriorly expanded girdle.

REPRODUCTION
Sexes are separate. Fertilization is external. Larvae have a free-swimming stage before settling to the ocean bottom.

CHITONIDA—CRYPTOPLACOIDEA—CRYPTOPLACIDAE
HIDDEN SHELL CHITONS

The family Cryptoplacidae contains a single genus *Cryptoplax* with more than twenty species. This genus occurs in shallow waters throughout the tropical and temperate regions of the Indian and western Pacific Oceans.

The animals are flexible, elongated, narrow, and worm-like, some reaching 6 in (150 mm) in length. The shell valves are longer than broad, reduced in size, and surrounded by a wide, fleshy girdle that is flexible, covered with spicules, and has tufts of bristles at the sutures. The shell valves are articulated in juveniles, but become separated as the animal grows, with the posterior-most valves being most widely separated. Insertion plates and sutural plates of the valves are extended anteriorly, and are sharp and smooth. The gills are limited to the posterior portion of the pallial groove.

Most species live on reefs and on and under rocks in shallow water. Their narrow, flexible bodies allow them to live in crevices and holes to which they can easily conform. Some species live on brown algae in the intertidal zone. Fossils found in Europe and Australia are known from the Miocene (23–2.6 mya).

DISTRIBUTION
Most species occur in temperate and tropical portions of the Indian Ocean and western Pacific Ocean and the Red Sea.

DIVERSITY
Family includes 18 living species assigned to the genus *Cryptoplax*.

HABITAT
On or under rocks and reefs. Some live in holes and crevices.

SIZE
Most species range in size from 2–6 in (50–150 mm).

DIET
Grazing omnivores, feeding on algae, sponges, and encrusting organisms.

REPRODUCTION
Sexes are separate. Gametes are released into the water column. Larvae have a free-swimming stage before settling to the ocean bottom.

ABOVE | A specimen of *Cryptoplax striata* on subtidal rock off southern Australia.

RIGHT | *Cryptoplax larvaeformis* under a rock in Bali, Indonesia. The shell plates are greatly reduced and enveloped in the muscular girdle, with only portions of three plates visible.

OPPOSITE | *Cryptoplax caledonicus* lateral and dorsal views of a preserved animal from New Caledonia with shell plates more widely separated posteriorly.

CHITONIDA—MOPALIOIDEA—MOPALIIDAE
SPECKLED CHITONS

Living representatives of the Mopaliidae family include more than seventy species that have distributions ranging from the Indo-Pacific to northern and northwestern Pacific and North Atlantic Oceans, as well as southern South America and the Southern Ocean off Antarctica.

Mopaliidae species are broadly ovate. Its shells have a tegmentum that is sculptured with the central and lateral areas of each valve differentiated. Sutural lamina and insertion plates are well-developed. The girdle may have bristles and spines but not scales. Gills extend the length of the pallial grooves.

Diets vary among lineages. Some taxa feed on algal films and coralline algae, while others are opportunistic grazing omnivores. Several are specialized predators of small crustaceans and worms captured beneath an expanded anterior girdle flap.

Members of the extinct subfamily Heterochitoninae are known from the Jurassic (200 mya) in Europe, while Mopaliinae first appear during the Miocene (25 mya) in New Zealand.

LEFT | *Mopalia lignosa* from intertidal rocks along the coast of Washington State.

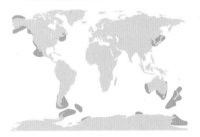

DISTRIBUTION
Global, primarily from temperate and polar regions, from intertidal to bathyal depths.

DIVERSITY
Family includes approximately 70 living species assigned to 10 genera within 1 subfamily.

HABITAT
Most species live on rocks and hard substrates in intertidal areas, some at depths to 3,280 ft (1,000 m).

SIZE
Most species range in size from 1–2 in (25–50 mm), some reach 5 in (125 mm).

DIET
Grazers, feeing on algae, or on a variety of animal material such as bryozoans,

LEFT | *Mopalia spectabilis* on rock in Browning Pass, Queen Charlotte Strait, British Columbia, Canada.

BELOW | Anterior view of *Placiphorella rufa* at a depth of 65 ft (20 m) in ambush position with raised anterior girdle flap to capture prey—off Irishman's Hat, Unalaska Island, Aleutian Islands, Alaska.

hydroids, and worms. Some species feed on sponges, while others capture small crustaceans and worms.

REPRODUCTION
Sexes are separate. Most species shed eggs and sperm into the water. Females of some species brood eggs and juveniles within the pallial groove.

MONOPLACOPHORA

Monoplacophorans have small, limpet-like (cap-like) shells and are represented in the fossil record since the earliest Cambrian (540 mya). They were common and diverse components of the shallow-water faunas of the early Paleozoic. They were thought to have become extinct during the Devonian (359 mya) until live specimens of Neopilinidae—the single living family—were collected at abyssal depths along the eastern Pacific Ocean in the 1950s.

Monoplacophora are considered to be at the base of Conchifera, which includes all mollusks with a true dorsal shell, and the sister group to the remaining conchiferan classes (Bivalvia, Scaphopoda, Gastropoda, and Cephalopoda).

SHELL AND ANATOMICAL FEATURES OF LIVING MONOPLACOPHORA

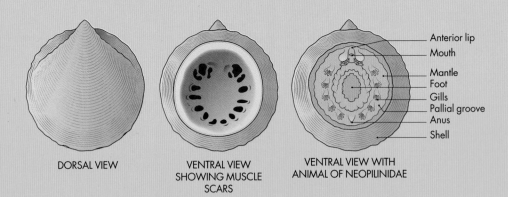

DORSAL VIEW

VENTRAL VIEW SHOWING MUSCLE SCARS

VENTRAL VIEW WITH ANIMAL OF NEOPILINIDAE

- Anterior lip
- Mouth
- Mantle
- Foot
- Gills
- Pallial groove
- Anus
- Shell

NEOPILINIDA—NEOPILINOIDEA—NEOPILINIDAE
NEW PILINA

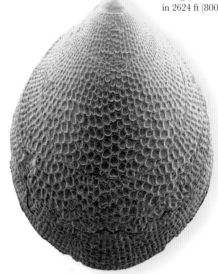

BELOW | *Micropilina wareni* from the West Norfolk Ridge, West of Cape Reinga, New Zealand, in 2624 ft (800 m).

The family Neopilinidae, which is presently known from 35 living species, is the only family within the order Neopilinida. The remaining four orders within the Class Monoplacophora became extinct by the end of the Devonian (349 mya). Since the discovery of the first living species of Neopilinidae at abyssal depths, additional species have been found in shallower waters, with shallow water species being smaller.

All species of Neopilinidae have shells that are ovate in outline, arched in profile, with the apex slightly projecting over the anterior margin. The shell exterior may have radial ribs and co-marginal growth lines that vary in prominence among species. The protoconch is spirally coiled in some genera. The shell interior has eight paired muscle scars. The animal has a large oval foot that is encircled by a pallial groove that contains serially repeated external gills and nephridial (kidney) openings. The head is anterior and lacks eyes or tentacles. The mouth is ventral, with anterior and posterior lips, leading to a long, looped alimentary system terminating in an anus along the posterior midline.

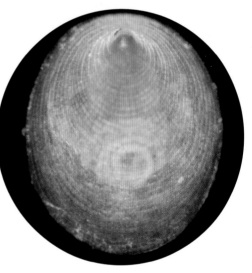

RIGHT | *Veleropilina oligotropha* from 680 miles (1,100 km) north of Hawai'i in 20,000 ft (6090 m).

DISTRIBUTION
Global, at depths ranging from 570–21,325 ft (174–6,500 m).

DIVERSITY
Family includes 7 genera.

HABITAT
Some species occur on hard substrates including manganese nodules; others live in soft sediments.

SIZE
Species range in size from ¼–1½ in (6–37 mm).

DIET
Deposit feeders, including diatoms, foraminiferans, and other small organisms.

REPRODUCTION
Sexes are separate, with gonads opening through the third and fourth nephridia (kidneys). Fertilization is external.

OPPOSITE | *Mytilus edulis*, White Sea, Russia.

BIVALVIA

Bivalvia, the second-largest class within the phylum Mollusca, is both diverse and abundant, with nearly 50,000 living species inhabiting a wide variety of marine and freshwater habitats. Bivalves are readily recognized by their shells, which are composed of two laterally compressed valves (left and right) connected by an elastic chitinous ligament along their dorsal midline, and a hinge that is composed of interlocking teeth and sockets that maintains alignment between the valves when they open and close. In most species the valves are mirror images of each other and capable of closing tightly when adductor muscles are contracted, and opening due to tension in the ligament when these muscles are relaxed. The shell valves are composed of an outer proteinaceous layer, the periostracum, and several layers of calcium carbonate with crystalline structure that varies among lineages.

Sculptural elements on the outer surfaces generally include concentric growth striations, with some families having pronounced axial ribs, or spines. The inner surfaces of the valves have scars indicating attachment areas of various muscles. The largest are the scars of the adductor muscles. A pallial line connecting the adductor muscle scars is where the muscular edge of the mantle connects to the shell. It may have an indented area called the pallial sinus, into which siphons of species that have them can be withdrawn.

The mantle of bivalves, the organ that secretes the shell in all mollusks, completely surrounds and encloses the entire body, forming a capacious mantle cavity that contains large gills and a muscular foot. Animals are bilaterally symmetrical, with an anterior mouth and a posterior anus, but lack a distinct head and associated organs like tentacles and eyes. Bivalves are the only mollusks that lack a radula.

The majority of bivalves are infaunal, living burrowed in sand, mud, or clay with only their siphons or mantle edges exposed. Others are epifaunal, and are attached by byssal threads secreted by their foot, or permanently cemented to rocks, or other hard substrates. Several are free living and capable of swimming short distances to avoid predators. Most bivalves filter food particles from the water using their gills. Some host algae or bacteria in their tissues that contribute to their nutrition. Others burrow in wood and can digest cellulose.

Living bivalves vary greatly in size. The smallest-known species *Condylonucula maya* grows to $1/_{32}$ in (0.5 mm) in size, while some giant clams (*Tridacna*) can grow to more than 5 ft (1.5 m) in length and weigh over 500 pounds (225 kg). Members of the extinct family Inoceramidae, which were abundant in the late Cretaceous, reached 6–9 ft (2–3 m) in length.

Although the oldest fossils of Bivalvia are from the early Cambrian (520–513 mya), they are next represented in the early Ordovician (470 mya), when their diversity began to increase. Many bivalve lineages became extinct at the end Cretaceous mass extinction, but surviving lineages have continued to diversify throughout the Cenozoic.

SHELL FEATURES AND ANATOMY OF THE VENERID BIVALVE *MERCENARIA*

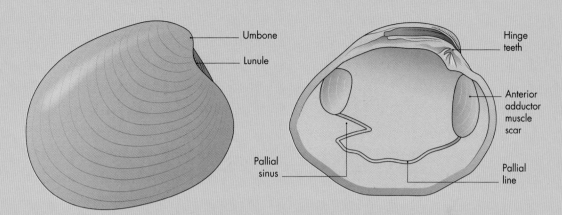

EXTERNAL VIEW OF RIGHT SHELL VALVE

INTERNAL VIEW OF LEFT SHELL VALVE

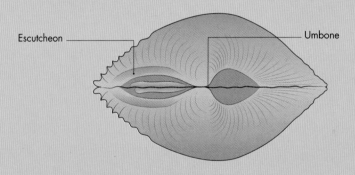

DORSAL VIEW OF SHELL VALVES

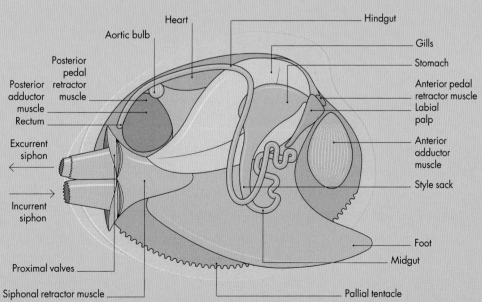

BIVALVE WITH RIGHT VALVE REMOVED TO EXPOSE ANATOMY

NUCULIDA—NUCULOIDEA—NUCULIDAE
NUT SHELLS

The family Nuculidae includes the smallest living bivalve ($1/32$ in / 0.5 mm) as well as other species that may reach 2 in (50 mm) in length. Nuculids are small, shallow burrowing marine bivalves that range from shallow-water to abyssal depths throughout all the oceans of the world. They live in soft mud to coarse sand, using their broad foot to move through the sediments and extending their palp proboscides to bring detritus from below the sediment surface into the shell and form food strands that are ingested. Nuculids are also capable of suspension feeding, especially as juveniles.

Nuculid shells are obliquely ovate to triangular in shape, bilaterally symmetrical, equivalve, moderately inflated, thin, and have a thick, smooth, yellowish-brown periostracum and a nacreous inner layer. The hinge is composed of multiple raised triangular teeth, separated by pits to receive teeth on the opposing valve. The ventral margin may be smooth or finely denticulate. Anterior and posterior

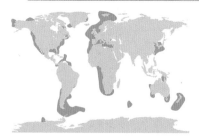

DISTRIBUTION
Cosmopolitan, ranging from sublittoral to abyssal depths.

DIVERSITY
Family includes approximately 200 living species assigned to 11 genera within 1 subfamily.

HABITAT
Burrowers in sediment; they remain close to the surface.

SIZE
Species range in size from $1/32$–$2\,3/4$ in (1–70 mm) in length.

DIET
Species of Nuculidae use their palp proboscides to move surface deposits into the mantle cavity and produce mucus food strands that are carried to the mouth.

OPPOSITE | *Nucula proxima* (shell about 5 mm / ¼ in)—from shallow muddy bottom off Sanibel Island, Florida.

ABOVE | *Nucula nucleus* on a sandy bottom in the Adriatic Sea.

REPRODUCTION
Sexes are separate. Fertilization is external. Larvae are lecithotrophic, being nourished by a large yolk. Larvae are of the pericalymma type. Some species have been reported to brood their young.

adductor muscles are similar in size and the pallial line is entire but weak. Animals have small gills with triangular filaments that are short and broad, large labial palps with long extensions (palp proboscides), a large foot with a broad sole, and lack siphons. Their blood contains the respiratory pigment hemocyanin.

With earliest records from the Ordovician (>440 mya), Nuculidae is the only living family in Nuculoidea.

SOLEMYIDA–SOLEMYOIDEA–SOLEMYIDAE
AWNING CLAMS

Members of the family Solemyidae can be recognized by their small, extremely thin, shells with a brown periostracum that extends beyond the calcified portions of the shell, giving rise to the common name Awning Clams.

The shells are elliptical, with dorsal and ventral edges nearly parallel and the posterior margin more rounded than the anterior. The hinge, which lacks teeth, is closer to the anterior margin. There is a

BELOW | *Acharax clarificata*, a deep-water species from off the eastern coast of North Island, New Zealand.

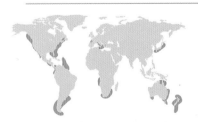

DISTRIBUTION
Occurs in all oceans except in the polar regions, ranging from subtidal to abyssal depths.

DIVERSITY
Family includes 32 living species assigned to 3 genera within 2 subfamilies.

HABITAT
U- or Y-shaped, mucus-lined burrows in anoxic mud bottoms. They can swim short distances by rapidly forcing water out of the mantle cavity with rapid, piston-like movements of their foot.

SIZE
Species range in size from ¼–4 in (6–100 mm) in length.

small rib on the inner surface just anterior to the posterior adductor muscle scar, which is smaller than the anterior adductor muscle scar.

Mantle margins are fused ventrally, with an anterior gape through which the large muscular foot may be extended, and a smaller posterior exhalant opening that is surrounded by small tentacles. The foot has a broad sole with thick papillae along its margins. The sole of the foot can be expanded laterally for burrowing into the sediment.

All species of Solemyidae host symbiotic, chemoautotrophic bacteria within the cells of their two bipectinate gills, on which they depend to varying degrees for their nutrition. As a consequence, the digestive tract is greatly reduced or even absent in some species. Hemoglobin has been identified in the gills of some solemyid species.

Earliest records of Solemyidae are from the Ordovician. During the Miocene, species nearly 12 in (300 mm) in length occurred in Japan. There are two subfamilies, three genera and thirty-two living species presently known. Of the two subfamilies, Solemyinae occur in shallow waters along continental shelf and upper continental slope depths ranging 3–1,970 ft (1–600 m), while members of the Acharacinae inhabit deep-water habitats 1,310–>16,400 ft (400–>5,000 m), including hydrothermal vents and methane seeps.

ABOVE | To propel itself, *Solemya velesiana* uses a large extendable foot fringed with tentacles—it can also swim by rapidly opening and closing its shell.

DIET
Solemyids host endosymbiotic chemosynthetic bacteria in their gills, from which they derive nutrition. They have a much reduced or absent gut and labial palps.

REPRODUCTION
Sexes are separate. Produce lecithotrophic pericalymma larvae that may have a brief planktonic phase before settling to the ocean bottom. There is no veliger stage.

NUCULANIDA—NUCULANOIDEA—YOLDIIDAE

YOLDIA CLAMS

Yoldiid bivalves are common members of muddy and sandy bottom faunas, and significant components of the diets of commercially harvested fish. Shells are of moderate size, elongated oval in shape, thin-walled,

BELOW | *Megayoldia thraciaeformis*, Cape Cod, Massachussetts, at a depth of 130 ft (40 m).

DISTRIBUTION
Global, with most species inhabiting temperate latitudes and polar regions, at depths ranging from subtidal to 16,400 ft (5,000 m).

DIVERSITY
Family includes 150 living species assigned to 8 genera within 2 subfamilies.

HABITAT
Burrows beneath the surface of sand and mud bottoms.

SIZE
Species range in size from ¼–2 in (2–60 mm).

DIET
Feed primarily on organic deposits in muddy and sandy bottoms but can supplement their diet by filter feeding.

ABOVE | *Yoldia limatula*, a species that inhabits temperate regions of the North Atlantic and North Pacific, revealing internal anatomy and extended foot.

laterally compressed, equivalved, and gaping at both ends. The anterior end is rounded; the posterior end may be slightly elongated. The periostracum is thin yellowish brown.

The shell is composed of aragonite and the interior is not nacreous. The hinge is broad, with interlocking raised triangular teeth on either side of the umbo. The anterior and posterior adductor muscle scars are nearly equal in size. The pallial line has a deep sinus. The mantle margins are not fused ventrally. The incurrent and excurrent siphons are long. The foot is large, with a broad, expandable sole with papillae along its margins. The gills are protobranch. Labial palps have long, ciliated palp proboscides.

The Yoldiidae live infaunally, buried beneath the surface of sand or mud bottoms. They are primarily deposit feeders that use their palp proboscides to excavate chambers adjacent to their burrows and are also capable of suspension feeding through their siphons. Yoldiids are rapid burrowers and contribute to bioturbation of the sediments in which they live.

The family Yoldiidae is well-represented in Cretaceous faunas.

REPRODUCTION
Sexes are separate. Larvae are pericalymma, with a large yolk, and are covered with a barrel-shaped epithelial layer that is discarded at metamorphosis.

ARCIDA—ARCOIDEA—ARCIDAE
ARK SHELLS

The shells of Arcidae are generally oval or trapezoidal in shape, slightly inflated, with valves equal in size or the left valve slightly larger. Some species have a gape in the ventral margin, through which the byssus is attached. Surface sculpture usually consists of weak radial and co-marginal ribs or ridges.

All members of the Arcidae have a characteristic long hinge that is straight or slightly curved, with a single row of numerous interlocking small teeth (taxodont dentition). Shells are covered with a thick fibrous periostracum that may have multiple hairlike margins but may be worn or abraded. The animals have two adductor muscles, the posterior generally larger. The mantle margins are not fused ventrally. Siphons are not present.

Small ocelli that are sensitive to light are present along the mantle margins of some species, with the animal reacting to shadows. Several species (e.g. *Tegillarca granosa*) contain hemoglobin as a respiratory pigment and have red blood cells in their tissues. Members of the Indo-Pacific genus *Trisidos* are

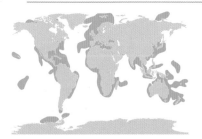

DISTRIBUTION
Temperate and tropical coastal areas worldwide, from intertidal zone to depths of about 150 ft (50 m).

DIVERSITY
Family includes approximately 300 living species assigned to 28 genera within 1 subfamily.

HABITAT
May be attached to rocks and hard substrate by a byssus, or burrow in sand with byssus attached to buried objects. Some species are capable of burrowing into rocks or coral.

SIZE
Species range in size from 1–6 in (25–150 mm).

BELOW | *Barbatia parvivillosa* on the underside of a rock in an intertidal pool at Mullaway Headland, New South Wales, Australia.

OPPOSITE | *Trisidos tortuosa*, a species common throughout the tropical Indo-West Pacific, has long byssal threads that anchor it in sandy bottoms.

DIET
Suspension feeders.

REPRODUCTION
Sexes are separate and fertilization is external. Larvae have a planktonic veliger stage. Some species have been reported to brood their young.

unusual in having a twisted shell, with the posterior end twisted up to 90 degrees relative to the anterior end of the shell. Some species of Arcidae have been reported to live for up to thirty years.

The family Arcidae had its origins in the Jurassic and is one of five families within Arcoidea. Recent molecular phylogenetic studies indicate that the family as currently understood may be paraphyletic.

MYTILIDA—MYTILOIDEA—MYTILIDAE
MUSSELS

With origins in the Devonian (358.9–419.2 mya), the family Mytilidae is one of the most diverse and species-rich families of Bivalvia. Shallow-water species such as the Blue Mussel (*Mytilus edulis*) and the Green Mussel (*Perna viridis*) are most familiar since they are harvested for human consumption. Other lineages have diversified to inhabit chemosynthetic communities along hydrothermal vents at abyssal depths (*Bathymodiolus*) and freshwater habitats (*Limnoperna*).

Shells tend to be elongated, inflated and equivalve, with the hinge near the narrow dorsal anterior end of the shell. The posterior ventral end of the shell is broader and rounded. The hinge dentition, when present, consists of small denticles. The shell's exterior may be smooth, have very fine commarginal growth striae, or radial ribs. It is covered by a brown to black periostracum that may be hirsute or calcified in some species.

The shell is composed of calcite, aragonite, or both, depending on species. The posterior adductor muscle is large, and the anterior adductor muscle is small or lost in some genera. The pallial line is entire and weak in most taxa, which lack siphons except for the subfamily Lithophaginae, which have siphons

LEFT | A cluster of mussels, *Musculus discors*, attached to a kelp stem, off Alaska.

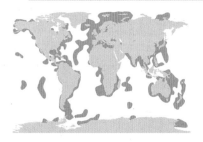

DISTRIBUTION
Global, from the intertidal zone to hydrothermal vents at abyssal depths. Several genera have become established in fresh water.

DIVERSITY
Family includes approximately 400 living species assigned to 59 genera within 12 subfamilies.

HABITAT
Many species are epifaunal, attaching to hard surfaces with their byssus and forming large and dense colonies. Other species may attach to stones that are partially buried in the sand. Members of the subfamily Lithophaginae burrow into corals or soft rock. Others (*Musculus*) may produce nests of byssal threads forming colonies. Others are endosymbionts of ascidians.

and a pallial sinus. The foot is long, finger-like, laterally compressed, and capable of rapid crawling. There is a byssal groove in the foot. The byssus is attached to large pedal retractor muscles that insert into the shell. Gills are filibranch.

Some mytilids (*Amygdalum*) have hemoglobin as a respiratory pigment.

ABOVE | *Bathymodiolus* mussels at the Menez Gwen hydrothermal vent off the Azores archipelago in the Mid-Atlantic (Portugal).

BELOW | Shell of the Common Blue Mussel *Mytilus edulis*.

SIZE
Species range in size from 1–8 in (25–200 mm).

DIET
Most species are suspension feeders. *Bathymodiolus* inhabits hydrothermal vents and is nourished by intracellular chemoautotrophic bacterial symbionts.

REPRODUCTION
Most species have separate sexes and are synchronous spawners, producing long-lived veliger larvae. Others brood.

OSTREIDA—OSTREOIDEA—OSTREIDAE
TRUE OYSTERS

Ostreidae, also known as oysters, are a diverse family of bivalves, many of which cement their shells to hard substrates and occur in dense populations, some giving rise to "oyster reefs" that provide a habitat for many other species. Some species produce shell outgrowths called "claspers" to attach to gorgonians or mangrove roots.

Ostreid shells are irregular in shape but tend to be rounded or elongated along the dorsoventral axis. The left valve, which is attached to the substrate, is concave and the right valve is nearly flat. The hinge lacks teeth in adults. Shells are calcitic. There is a single adductor muscle scar (posterior adductor) located near the center of the shell. The mantle edge is open, with the middle mantle fold forming a veil. The mantle is not fused nor are there siphons. The foot and byssus are absent in adults. The psuedolamellibranch gills are large and partially encircle the adductor muscle.

Sexes are separate, although some species are protandrous hermaphrodites. Others may be simultaneous hermaphrodites. Males shed sperm into the water. Depending on taxa, fertilization may be external or occur within the female, with eggs remaining in the mantle cavity until the prodisoconch stage.

The family Ostreidae arose during the Late Triassic. The subfamilies Ostreinae and Crassostreinae, which include several species harvested for human consumption, are more broadly distributed in temperate and tropical seas, while the subfamily Lophinae tends to be predominantly tropical.

DISTRIBUTION
Circumglobal, inhabiting all continents except Antarctica.

DIVERSITY
Family includes 74 living species assigned to 16 genera within 4 subfamilies.

HABITAT
Wide range of marine habitats, both tropical and temperate, fully marine and estuarine. Various species occur from the intertidal zone to depths of about 650 ft (200 m).

SIZE
Species range in size from 1–2 in (25–50 mm) to over 12 in (300 mm) in length.

ABOVE | Intertidal oyster reef of *Crassostrea virginica* off the coast of Charleston, South Carolina.

RIGHT | *Lopha cristagalli* covered with encrusting sponge on a subtidal reef in the Solomon Islands.

OPPOSITE | Two views of the shell of *Crassostrea virginica*, showing the outer surface of the upper (right) valve and the inner surface of the more convex lower (left) valve, which is attached to the substrate.

DIET
Filter feeders.

REPRODUCTION
Some species have separate sexes; others may be protandrous or simultaneous hermaphrodites. Fertilization may be external, or within the mantle cavity of females, which then brood the young.

OSTREIDA—PTERIOIDEA—MALLEIDAE
HAMMER OYSTERS

Members of the family Malleidae are referred to as Hammer Oysters because the shells are shaped like a hammer or the letter T to varying degrees, depending on species. The shell is large, laterally compressed, dorsoventrally elongated, and irregular in shape, due primarily to frequent breakage and repair. Both anterior and posterior auricles of the hinge become elongated with the umbones near the midpoint. The ventral margin is narrowly elongated to form the "handle of the hammer." The shell has a posterior gape. The outer layer of the shell is composed of calcite, with an aragonitic nacreous inner layer limited to the region around the single adductor muscle scar and visceral mass. There is no pallial line or sinus.

The animal has a single large posterior adductor muscle, and a long tubular foot with a byssal gland at one end and a longer flexible end that is used to

DISTRIBUTION
Temperate and tropical seas worldwide at depths ranging from subtidal to 330 ft (100 m).

DIVERSITY
Family includes 9 living species assigned to 2 genera within 1 subfamily.

HABITAT
Some species are attached to rocks or other hard objects by byssus; others settle in crevices between rocks and rubble. Some lie on the surface of sand or mud bottoms among seagrasses. Several species tend to occur in large colonies.

SIZE
Species range in size from 2–12 in (50–300 mm).

clean the inhalant chamber of the mantle cavity. Siphons are not present. Mantle margins are not fused, but inner mantle margins form a pallial veil. Long fillibranch gills extend the length of the mantle cavity. The mantle is capable of rapid shell repair. Cephalic eyes are present in most species, and some also have pallial eyes.

Malleidae occur in temperate and tropical seas throughout the world; some in large colonies or co-occurring with species of Isognomonidae. They are marine suspension feeders, with a diet rich in diatoms and dinoflagellates. Early records of Malleidae date to the Jurassic.

BELOW | Inner and outer views of the shell of *Malleus albus*.

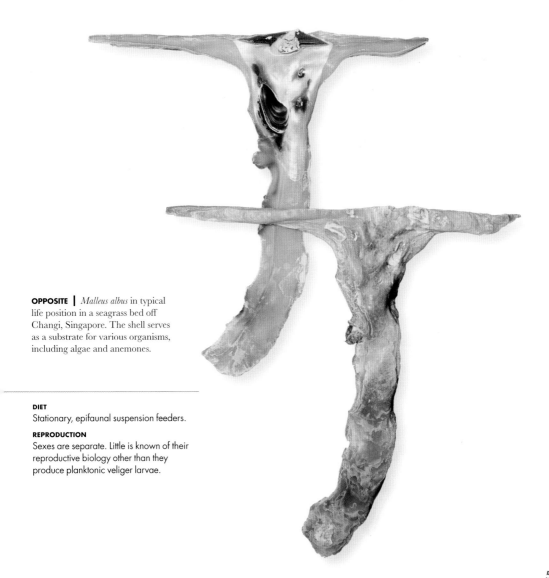

OPPOSITE | *Malleus albus* in typical life position in a seagrass bed off Changi, Singapore. The shell serves as a substrate for various organisms, including algae and anemones.

DIET
Stationary, epifaunal suspension feeders.

REPRODUCTION
Sexes are separate. Little is known of their reproductive biology other than they produce planktonic veliger larvae.

OSTREIDA—PTERIOIDEA—PTERIIDAE
PEARL OYSTERS, WING OYSTERS

Pteriidae inhabit tropical and subtropical shallow marine habitats, attaching to a variety of hard substrates by byssal threads.

Shells may be equivalve or inequivalve, are laterally compressed, obliquely ovate to circular, and have a straight hinge with short anterior and in some taxa long posterior auricles that appear as wing-like projections. The hinge lacks teeth in adult specimens. The right valve has a deep byssal notch. The exterior sculpture consists of fine, commarginal growth lines and periostracum that may become worn with age. The shell interior has a nacreous inner layer (mother of pearl) that may vary in color from white to pink, yellow, or black. A single (posterior) adductor muscle is present and is partially encircled by large gills. The animal lacks siphons. The foot is short and has a byssal groove. The mantle margins are not fused, and some species may have simple eyes along the mantle edge.

Species of the genera *Pteria* and *Pinctada* have been the sources of pearls through much of human history. Several Indo-Pacific species of *Pinctada* are farmed and used to produce cultured pearls. Among these is *Pinctada margaritifera*, which is the source of black pearls cultured in eastern Polynesia. Although the Atlantic species *Pinctada imbricata* had been the primary source of natural pearls for Europe during

DISTRIBUTION
Worldwide in tropical and subtropical seas, mostly along continental shelves, at depths ranging from subtidal to about 260 ft (80 m).

DIVERSITY
Family includes approximately 60 living species assigned to 3 genera within 1 subfamily.

HABITAT
Epifaunal and attached to hard substrates by a byssus, ranging from rocks and corals to gorgonians. Many species occur in clusters of multiple individuals and, in some cases, form banks along shallow coasts.

SIZE
Most species range in size from 1½–4 in (40–100 mm), with some reaching 12 in (300 mm) in length.

the Renaissance, the pearls it produces are smaller and often yellowish and it is no longer commercially harvested.

The oldest fossils of the family Pteriidae date to the Triassic (about 230 mya). Some recent studies have separated the genus *Pinctada* into a separate family, the Margaritidae.

LEFT | A young specimen of *Pteria hirundo* collected off the coast of Greece, showing the periostracum.

OPPOSITE | *Pinctada margaritifera* attached by its byssus. Mantle margins and a portion of the gill are visible.

RIGHT | Two valves of *Pinctada margaritifera*. The inner surface is covered with mother of pearl and shows a pearl near the adductor muscle scar. The outer surface is worn and encrusted.

DIET
Feed on fine suspended particles that they filter from the surrounding water with their gills.

REPRODUCTION
Protandrous hermaphrodites, initially developing as males and becoming females after several reproductive cycles. Fertilization is external. Larvae develop in the plankton for 3–4 weeks before settling as spat.

OSTREIDA—PINNOIDEA—PINNIDAE
PEN SHELLS

Pinnidae may be recognized by their triangular shell shape with flattened valves that are equal in shape and size, with a long, straight hinge line. The shell is thin and may be somewhat flexible. There is a gape along the anterior end of the ventral margin through which the byssus extends. The sculpture may be smooth to radially ribbed, often with scales that are inrolled and appear tubular. The anterior adductor muscle is small and located at the anterior margin of the shell, while the posterior adductor muscle is much larger and situated near the central part of the shell. The nacreous layers only extend as far as the posterior adductor muscle.

DISTRIBUTION
Cosmopolitan, inhabiting intertidal to subtidal depths in tropical and subtropical seas.

DIVERSITY
Family includes approximately 50 living species assigned to 3 genera within 1 subfamily.

HABITAT
Partially to almost completely buried in sediment, with the pointed anterior end downward, anchored to small stones or shell fragments by long byssal threads. The broad posterior end is at or above the surface of the sand or mud.

SIZE
Species range in size from 6 in (150 mm) to nearly 40 in (1 m) in length.

RIGHT | *Pinna nobilis* found in shallow waters off Sicily, Italy.

BELOW | The byssus of *Pinna nobilis*.

OPPOSITE | A group of *Pinna nobilis* in a shallow-water seagrass bed in the Mediterranean.

DIET
Suspension feeders, filtering particles with their gills.

REPRODUCTION
Sexes are separate, but some species may be hermaphroditic, producing planktonic veliger larvae.

The presence of a specialized pallial organ, a stalk-like feature with a distal glandular region situated dorsal and posterior to the posterior adductor muscle, is a feature unique to Pinnidae that is used to keep the mantle cavity clear and repair damage to the exposed posterior region of the shell. Pea crabs (family Pinnotheridae) may inhabit the mantle cavity of pinnids in a commensal relationship (where one organism benefits and the other is unharmed and receives no benefit).

The family Pinnidae dates to the Carboniferous (359–299 mya). *Pinna nobilis*, the largest species of Pinnidae and among the largest species of Bivalvia, may approach 40 in (1 m) in shell length. Its large byssus, known as sea silk, has been harvested since Roman times and woven into various items. It is thought that Jason's "Golden Fleece" was woven from the byssus of *Pinna nobilis*. This species, now protected, is endemic to the Mediterranean Sea and is now considered to be endangered.

LIMIDA—LIMOIDEA—LIMIDAE
FILE CLAMS

The obliquely elongated, equivalved shells of Limidae, which in many species have radial ribs with multiple scales, resemble a file or rasp. The shells are compressed, tend to have a narrow gape along their anterior margin, a hinge of two small auricles, and a triangular cardinal area with a shallow ligament pit in both valves. The animal has a single adductor muscle (anterior adductor muscle is absent) and a mantle that may be brightly colored (red or orange) with multiple, long tentacles projecting from the edge. Genera that inhabit shallower waters (*Lima, Ctenoides*) have pallial eyes along the mantle edge while those that live in deeper water in the aphotic zone lack eyes. Unlike other bivalves, the foot of limids is rotated 180 degrees, relative to the visceral mass. Limidae have defensive tentacles capable of autotomy (can detach) and secrete predator-deterring mucus.

Some species are attached to the substrate by byssal threads, which are shed to escape predators. Some live in aggregations in nests made of byssal fibers and mucus. Others live under rocks or corals and can swim short to moderate distances by rapidly opening and closing their valves, often accompanied by a rowing action of the tentacles.

LEFT | *Ctenoides ales*, called the Electric File Clam, is capable of a flashing light display by exposing highly reflective tissue along its mantle margin. It occurs throughout the tropical central Indo-Pacific region.

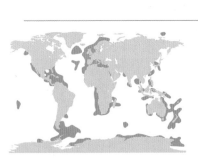

DISTRIBUTION
Cosmopolitan, inhabiting shallow waters around reefs at tropical and temperate latitudes. Others occur along the continental slope in colder water habitats.

DIVERSITY
Family includes approximately 200 living species assigned to 10 genera within 1 subfamily.

HABITAT
Shallow-water species live on reefs or under rocks and in crevices. Colder water genera extend from the Arctic to Antarctic waters and occur at depths as great as 10,500 ft (3,200 m).

SIZE
Species range in size from ¼–8 in (6–200 mm). Most shallow-water species have shells 2–3 in (50–75 mm) in height, while some deep water species (*Acesta*) may reach 8 in (200 mm).

LEFT | *Mantellina translucens* attached by byssus to rocky wall in 951 ft (290 m) off Willemstad, Curaçao in the Caribbean. The gills (ctenidia) are extended beyond the shell margins to filter food particles from the water.

BELOW | *Acesta bullisi* collected at a depth of 1,740 ft (530 m) on Viosca Knoll, south of Pascagoula, Mississippi.

Other genera (*Mantellina, Acesta*) inhabit deeper waters and tend to have thin or translucent shells. Shells of thin-shelled species may have multiple-repaired predator-induced fractures that suggest chemical defense mechanisms.

The order Limida had its origins in the Carboniferous (359–299 mya).

DIET
Filter feeders, capturing food particles in their gills.

REPRODUCTION
Sexes are separate, although some species have been reported to be capable of changing from male to female. Most species produce planktonic veliger larvae, while some brood their young in the suprabranchial chamber.

PECTINIDA—PECTINOIDEA—PECTINIDAE
TRUE SCALLOPS

Pectinids are commonly referred to as Scallops or True Scallops to distinguish them from the closely related Glass Scallops (family Propeamusiidae). Some can swim short distances by rapidly opening and closing their valves; this expels water, which propels them.

Readily recognized by their round-to-triangular fan-like shape with prominent, usually asymmetrical auricles (wings) flanking the umbo, a True Scallop's anterior auricle is usually larger. A byssal notch is present on the right valve below the auricle and lined with a comb-like row of denticles called a ctenolium that serves to separate the byssal threads. The ctenolium is a diagnostic feature of Pectinidae.

Shells are equivalve, thin and relatively flat, or, asymmetrical in species that lie on the bottom, with the right (lower) valve more convex and the left

BELOW | *Argopecten irradians* resting in a seagrass bed in shallow water. Pallial tentacles and blue pallial eyes are visible along the mantle edges.

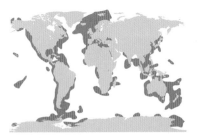

DISTRIBUTION
Global, from the tropics to the polar regions and from intertidal to hadal depths (0–23,000 ft / 0–7,000 m).

DIVERSITY
Family includes approximately 400 living species assigned to 65 genera within 4 subfamilies.

HABITAT
Epifaunal. Most are attached to hard substrates by a byssus, but can detach to avoid predators. Several are permanently cemented to rocks. Others are free-living on soft bottoms or in seagrass beds.

SIZE
Free-living species range in size from ½–12 in (12–300 mm). Species cemented to the substrate may reach 25 in (625 mm).

BELOW | *Gloripallium speciosum* is a species that is byssally attached to the underside of rocks and dead coral throughout the tropical western Pacific Ocean.

(upper) valve flatter or concave. The surface may be nearly smooth in some species, but most have radial ribs of varying prominence that interdigitate along the valve margins. Ribs may be ornamented with knobs or scales. Their aragonite and calcite shells are often brightly colored. The thin periostracum is usually eroded. There is a single large posterior adductor muscle scar near the shell's center. The mantle lobes are not fused ventrally and siphons are absent. The small foot has a byssal groove that secretes the byssus. Large gills nearly encircle the adductor muscle. Mantle edges have tentacles and eyes (generally blue) capable of some degree of image formation.

Pectinidae's fossil record extends to the Triassic. Scallop images have featured prominently in art and heraldry throughout the ages. The St. James Scallop (*Pecten jacobaeus*) has been used since medieval times to identify pilgrims on the Way of St. James.

RIGHT | *Pecten jacobaeus*, the St. James's Scallop or Pilgrim's Scallop, is a common offshore species living throughout the Mediterranean and along the eastern Atlantic from Portugal to Angola.

DIET
Suspension feeders

REPRODUCTION
Sexes are separate in some species; others are protandric hermaphrodites. Fertilization is external, with eggs developing into planktonic larvae.

PECTINIDA—PECTINOIDA—SPONDYLIDAE
THORNY OYSTERS

Members of the family Spondylidae are epifaunal bivalves that are readily recognized by having an inflated, roundish shell that is cemented to the substrate (usually rocks or coral reefs) by its right valve as well as by its surface sculpture of radial ribs or rows of spines most prominent on the left (upper) valve. The number and size of the spines varies among species, but longer spines tend to occur in calm waters.

Both valves have small anterior and posterior auricles. The hinge is straight and strong, with interlocking "ball and socket" teeth that maintain

RIGHT | *Spondylus americanus* collected on rubble in 150 ft (50 m) of water off Fort Lauderdale, Florida.

DISTRIBUTION
Tropical coastal areas worldwide, from the intertidal zone to depths of around 150 ft (50 m).

DIVERSITY
Family includes approximately 70 living species assigned to 1 genus.

HABITAT
Most species cemented to rocks and reefs at subtidal depths. Also found on sunken wrecks and offshore oil platforms.

SIZE
Species range in size from 1–6 in (25–150 mm) but may reach over 9 in (230 mm), depending on the length of spines.

DIET
Filter feeders, consuming plankton filtered by its gills.

REPRODUCTION
Sexes are separate; fertilization is external. Larvae have a planktonic stage.

ABOVE | *Spondylus varius* on reef, off the island of Mayotte in the Indian Ocean. The inner mantle folds (blue) and middle mantle folds (with red tentacles) are clearly visible.

the alignment of the valves and are characteristic of the family.

Many species come in a variety of bright colors, including red, orange, and yellow. The animal has a single, large adductor muscle located near the center of the shell. The large gills lie along the adductor muscle. The foot is used to clean the inhalant region of the mantle cavity. The mantle lacks siphons. The inner mantle folds form a veil with colorful patterns. The middle mantle folds have short tentacles and pallial eyes that react to shadows.

The fossil record of Spondylidae extends to the Jurassic. Species of *Spondylus* inhabit primarily tropical coastal areas, usually in areas with moderate currents, but without waves. The spines on the shell are usually overgrown by sponges and other epibionts that help camouflage the animal.

PECTINIDA—ANOMIOIDEA—ANOMIIDAE
JINGLE SHELLS

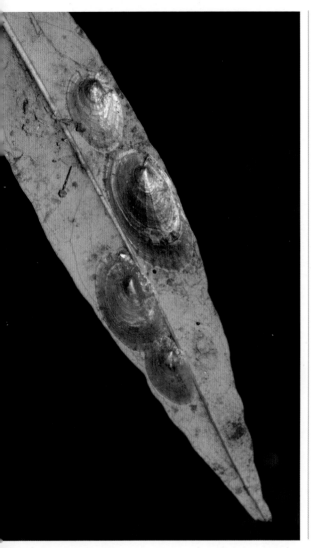

Shells of the family Anomiidae are circular to oval, inequivalve, thin-walled, and translucent, with the bottom (right) valve conforming to the topology of the substrate to which they are attached. The bottom valve has a large circular opening below the umbone, through which it attaches to the substrate by its byssus, which may become calcified. The upper (left) valve is more convex and conforms to the outline of the bottom valve. Sculpture consists primarily of irregular commarginal growth bands, although some species may have radial ribs. Shells are composed of a calcitic outer layer and an aragonitic inner layer that does not extend beyond the pallial line.

This animal has only a posterior adductor muscle. Siphons are absent, and mantle margins are not fused but form a pallial veil. The foot is slender and used to clean the area around the byssus. The filibranch gills partially encircle the byssus and posterior adductor muscle.

Anomiids are epifaunal, inhabiting temperate and tropical seas and estuaries throughout the world, primarily in intertidal to subtidal depths. Most are permanently attached to hard substrates by their byssus. A single species, *Enigmonia aenigmata*, is mobile

LEFT | *Enigmonia aenigmatica*, attached to a mangrove leaf along the coast of Singapore.

DISTRIBUTION
Temperate and tropical waters worldwide.

DIVERSITY
Family includes approximately 30 living species assigned to 9 genera within 1 subfamily.

HABITAT
Attached to hard substrates such as rocks, docks, or other shells by a byssus that passes through the hole in the bottom valve. Intertidal to around 330 ft (100 m). More common in areas with low current flow.

SIZE
Species range in size from 1–5 in (25–125 mm).

DIET
Filter feeders.

REPRODUCTION
Sexes are separate. Eggs develop into planktonic veliger larvae.

as a juvenile, capable of detaching and crawling or drifting before reattaching in mangroves, but becomes sessile as an adult. Anomiids are suspension feeders, filtering seawater through their gills when submerged.

The oldest records of the family Anomiidae are from the Jurassic.

ABOVE | *Pododesmus macrochisma*, attached to a dock in Puget Sound, Washington State.

BELOW | *Anomia simplex*—a common species that comes in a variety of colors, from yellow to white to gray/black.

TRIGONIIDA—TRIGONIOIDEA—TRIGONIIDAE
BROOCH SHELLS

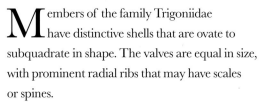

BELOW | *Neotrigonia margaritacea*—showing complex schizodont hinge structure.

Members of the family Trigoniidae have distinctive shells that are ovate to subquadrate in shape. The valves are equal in size, with prominent radial ribs that may have scales or spines.

The shell is aragonitic, with the exterior covered by a brownish periostracum and the interior nacreous, with radial grooves that correspond to the external ribs. The large schizodont hinge, which can occupy one-third of the shell volume, is diagnostic of Trigoniidae. It is composed of large teeth, each with raised ridges and grooves that interlock with those on the opposing valve.

The posterior adductor muscle is slightly larger than the anterior adductor muscle. The mantle is not fused. Gills are filibranch, with mineralized support along the length of the gill filaments.

The animals lack siphons. The foot is large and muscular with papillae along its posterior end.

Trigoniids are rapid and active burrowers within sediments, with the posterior-most portion of their shells remaining above the surface, enabling them to filter feed.

The earliest records of the order Trigoniida are from the Silurian (>420 mya), with a major

DISTRIBUTION
Australian coast.

DIVERSITY
Family includes 8 living species assigned to 1 genus.

HABITAT
Burrowers, with the posterior portion of the shell oriented above the surface of the sediment. Subtidal to depths of 1,310 ft (400 m), most commonly around 10–330 ft (3–100 m).

SIZE
Species range in size from 1–2 in (25–50 mm).

DIET
Filter feeders.

REPRODUCTION
Sexes are separate. Eggs have large yolks. At least one species has non-planktotrophic larvae.

radiation during the Jurassic (140 mya) that gave rise to dozens of families with hundreds of species and a worldwide distribution. Nearly all became extinct at the end of the Cretaceous. Only the family Trigoniidae, which is restricted to the coasts of Australia, survives in the recent fauna.

ABOVE | *Neotrigonia lamarckii*, a species endemic to shallow waters along southeastern Australia, with foot extended.

CARDITIDA—CARDITOIDEA—CARDITIDAE
CARDITA CLAMS

Carditidae are a large group of mobile, filter-feeding bivalves tracing their origins to the Devonian, but which diversified during the late Cretaceous and Cenozoic. Most species inhabit shallow waters in tropical and temperate latitudes throughout the world, and are byssally attached to rocks or live in crevices.

Carditidae shells are small to medium sized, inflated, and oval to rounded trapezoidal in outline. The valves are equal in size, solid, with strong radial sculpture that may have nodules or scales on the ribs, and crenulated margins corresponding to the ribs. The periostracum may be thick and dark brown. Shells are composed of aragonite. Anterior and

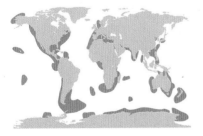

DISTRIBUTION
Shallow waters along continental shelves in tropical and temperate latitudes. Several small species occur at depths up to 4,920 ft (1500 m).

DIVERSITY
Family includes approximately 215 living species assigned to 39 genera within 7 subfamilies.

HABITAT
Many species are byssally attached under rocks, in crevices, and/or on gravel bottoms. Some are shallow burrowers in soft substrates.

SIZE
Species range in size from ¼–4 in (6–100 mm).

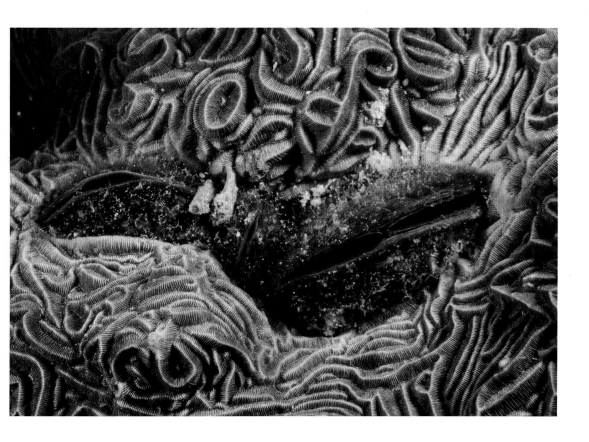

OPPOSITE | *Cardiocardita gabonensis*, found off the coast of Gabon, West Africa.

ABOVE | Three *Beguina semiorbiculata* within a crevice in a coral in Philippine waters.

DIET
Suspension feeders.

REPRODUCTION
Sexes are separate. Females produce few large, yolky eggs. Fertilization is within the female mantle cavity. Some species have planktonic larvae that feed on the yolk; others brood their young, releasing benthic juveniles.

posterior adductor muscle scars are usually equal in size, although the anterior muscle scar may be smaller in some species with a more elongated shell. The pallial line is entire, without a pallial sinus. The hinge is wide with prominent teeth.

Siphons are absent. Mantle folds are fused to form a posterior exhalant aperture, with the inhalant aperture only temporarily defined by apposition of inner mantle folds. Eulamellibranchiate gills are large, with demibranchs not equal in size. Blood of at least some species contains hemoglobin.

ANOMALODESMATA—CLAVAGELLOIDEA—PENICILLIDAE
WATERING POT SHELLS

Species of the family Penicillidae begin life with a bivalved shell. They may attach to or burrow into soft rock or into sand or mud bottoms and incorporate their juvenile, bivalved shell (about $1/8$ in / 3 mm in length) into the dorsal, anterior region of a much larger calcareous tube that is oriented vertically. The anterior of the tube remains buried and is expanded into a sieve-like structure that includes numerous small holes or tubules through which water flows. The posterior region is narrower and extends above the surface of the substrate where it may be covered with encrusting organisms.

In most genera, the animals are weakly attached to the shell and the anterior and posterior adductor muscles may be present, reduced, or lost. The body is enclosed in a tubular muscular integument that extends posteriorly and contains the elongated gills. The very long gills are composed of two V-shaped demibranchs that are attached to the visceral mass and become fused within the posterior portion of the integument, separating it into incurrent and excurrent sections. The tubular integument terminates in short, separate inhalant and exhalant siphons that are fused.

The fossil record indicates that the Penicillidae arose during the Late Oligocene (33–23 mya) and expanded throughout the Indo-West Pacific.

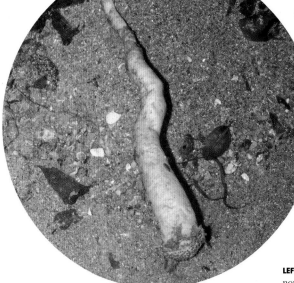

LEFT | *Verpa penis*, which is normally buried vertically in the sand, is shown on its side.

DISTRIBUTION
Temperate and tropical latitudes of the Indo-West Pacific.

DIVERSITY
Family includes 9 living species assigned to 6 genera within 1 subfamily.

HABITAT
Partially buried in sand among seagrass beds in shallow water.

SIZE
Tube length ranges in size from 3 in (75 mm) to over 16 in (400 mm).

DIET
Filter feeders.

REPRODUCTION
Species are hermaphroditic, but little is known about the development or reproductive biology of this family.

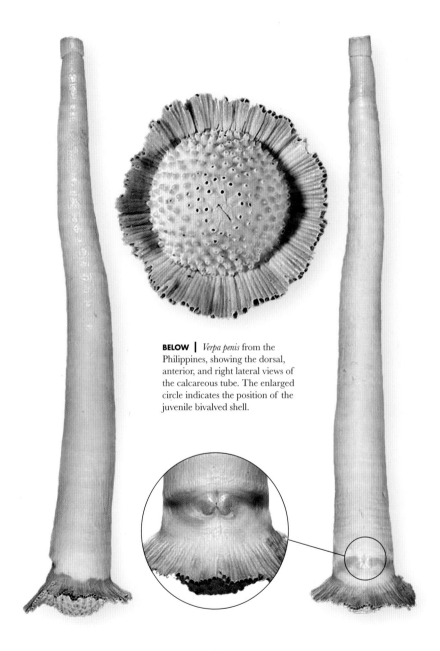

BELOW | *Verpa penis* from the Philippines, showing the dorsal, anterior, and right lateral views of the calcareous tube. The enlarged circle indicates the position of the juvenile bivalved shell.

ANOMALODESMATA—CUSPIDARIOIDEA—CUSPIDARIIDAE
DIPPER CLAMS

Cuspidariidae are carnivorous bivalves that live in sand and sediment throughout the world's oceans. Their shells are small, thin, whitish, and generally ovate to pyriform in shape, some with a long tube-like posterior rostrum. They are usually equivalve with an inflated anterior region. Sculpture may be smooth, or with strong radial ribs or concentric growth lines, depending on the genus and species. The hinge varies widely within the family. Shells are composed of aragonite and covered with a thin, yellowish periostracum. The pallial line is simple, with a weak pallial sinus.

The mantle is thin and fused, except for the siphonal aperture and pedal gape. The posterior adductor muscle is slightly smaller than the anterior adductor. The foot is long and tapered.

The siphons are asymmetrical, with an extremely expandable inhalant siphon being long and wide, and the exhalant siphon narrow. Both are surrounded by a siphonal sheath that secretes the rostrum. The inhalant siphon has four sensory

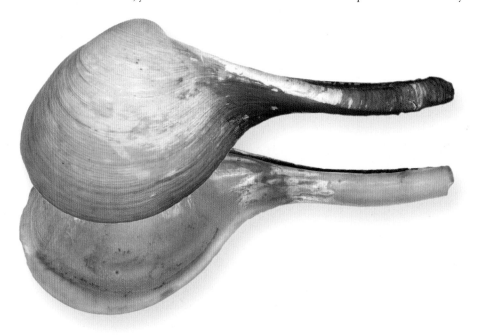

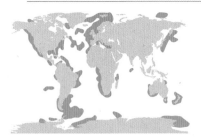

DISTRIBUTION
Cosmopolitan, occurring at all latitudes, primarily at bathyal and abyssal depths of 328 ft (100 m) to over 915 ft (3,000 m).

DIVERSITY
Family includes approximately 250 living species assigned to 19 genera within 1 subfamily.

HABITAT
Burrowers in soft sediments with the tips of the siphons exposed to detect prey.

SIZE
Species range from $5/16$– 2 in (8–50 mm) in size, some with a long rostrum that may account for the length of almost half of the shell.

ABOVE | *Cardiomya pectinata*, from off the coast of British Columbia, Canada.

OPPOSITE | *Cuspidaria rostrata*, a species with a long rostrum, from 390 ft (120 m) depth off Massachusetts.

DIET
Polychaetes (marine worms), foraminifera, small crustaceans, and other bottom-dwelling animals.

REPRODUCTION
Sexes are separate in studied species. Offspring develop as lecithotrophic larvae or undergo direct development.

tentacles at its tip; the exhalant siphon has three. The gills are septibranch, composed of a thick muscular septum with ciliated pores.

Cuspidariids are infaunal burrowers, with the opening of the siphons nearly level with the sediment surface. Small prey activate the sensory papillae on the siphon tips. They are then captured by expanding and extending the inhalant siphon and contracting the septum to suck the prey into the mantle cavity. The labial palps and foot then push the prey toward the wide mouth.

Oldest records of the family Cuspidariidae date to the Jurassic.

LUCINIDA—LUCINOIDEA—LUCINIDAE
LUCINA CLAMS

BELOW | *Lucina pensylvanica* from Jupiter Sound, Florida. Upper view shows the thin periostracum with calcified scales.

The family Lucinidae are among the most diverse groups of chemosymbiotic invertebrates. They rely on sulfur-oxidizing bacterial symbionts living within their gills that utilize hydrogen sulfide as an energy source to produce enough sugars to sustain themselves and their lucinid hosts.

Lucinids must acquire bacterial symbionts from their surrounding environment shortly after the larvae metamorphose and settle into the sediment.

Lucinid shells are generally small, solid, ovate, or round in outline, equivalve, moderately inflated, and without a gape. The umbone is centrally located and the hinge is heterodont. The shell exterior usually has pronounced commarginal sculpture and, in some cases, radial ribs that may produce a beaded pattern. The inner margins of the shell are smooth or finely denticulated. The anterior muscle scar is larger and more elongated than the rounded posterior muscle scar. The pallial line joining them is uninterrupted.

The animal lacks siphons. The mantle margins are not fused ventrally but are fused posteriorly to form an excurrent and usually an incurrent aperture. The foot is very long and extensible and is used to form a mucus-lined incurrent tube in the sediment. The large gills are thickened and harbor the

DISTRIBUTION
Nearly global distribution, with species occupying infaunal habitats from mangrove swamps and seagrass beds to hydrocarbon seeps at abyssal depths.

DIVERSITY
Family includes approximately 500 living species assigned to 96 genera within 8 subfamilies.

HABITAT
Deep burrowers into sulfide-rich mud, sand, or gravel sediments.

SIZE
Most species range in size from ¼–2 in (6–50 mm), with some reaching nearly 6 in (150 mm). Eocene fossils more than 12 in (300 mm) in height have been reported.

chemosynthetic bacteria within specialized bacteriocyte cells. The blood contains intracellular hemoglobin.

A single widespread bacterial species was found to be the most common symbiont in several different lucinid species inhabiting the Atlantic, Pacific, and Indian Oceans in both northern and southern hemispheres.

The family Lucinidae first appeared in the fossil record during the Silurian (333–417 mya) and diversified during the Upper Cretaceous.

ABOVE | *Codakia tigerina* is a wide-ranging shallow-water species that lives buried in sand near coral reefs throughout the tropical Indo-Pacific.

RIGHT | *Ctena decussata* is a species that often occurs in shallow-water seagrass beds in the Mediterranean Sea and along the temperate eastern Atlantic coast.

DIET
Sugars produced by sulfur-oxidizing symbiotic bacteria living within their gills.

REPRODUCTION
Sexes are separate. Fertilization is external, with planktonic veliger larvae.

ADAPEDONTA—HIATELLOIDEA—HIATELLIDAE
GEODUCKS

The Hiatellidae is a family that includes one of the largest burrowing bivalves (*Panopea generosa*), a popular commercially harvested species with a shell nearly 12 in (300 mm) in length and a siphon than can be extended 2 ft (0.6 m) beyond the shell. Depending on genus, shells are ovate to subquadrate in shape, with thick valves that are equal in size, and with widely gaping anterior and posterior ends. The umbone is centrally located with a simple hinge, which may be lost in large adults. The exterior sculpture may be smooth or have strong, irregular, commarginal growth lines that may be rugose.

The shell is chalky white in color, with a brown periostracum. Some of the smaller species may have radial ribs with spines. The interior is porcelaneous white. Both adductor muscles are similar in size. The pallial line is impressed and may be discontinuous. Mantle margins are fused, with a small pedal gape for the finger-like foot, which has a byssal groove. The incurrent siphon is larger than the excurrent siphon; both are muscular, partially united, and covered by periostracum. The siphon is too large to be retracted into the shell. Gills are eulamellibranchiate and may extend partially into the excurrent siphon.

DISTRIBUTION
Global, in a variety of environments, from polar regions to the deep sea.

DIVERSITY
Family includes 22 living species assigned to 4 genera within 1 subfamily.

HABITAT
Some burrow in sand to considerable depths; others nestle into crevices in rocks or bore into soft rocks.

SIZE
Species range in size from 1–12 in (25–300 mm), though they are typically 6–8 in (150–200 mm) and weigh about 7 lbs (3.2 kg). Siphons of some species can extend as much as 3 ft (0.9 m).

DIET
Suspension feeders.

OPPOSITE | Shell of *Panopea generosa* from off San Francisco, California.

ABOVE | A living *Panopea generosa* dug up in Puget Sound, Washington State.

REPRODUCTION

Sexes are separate and fertilization is external, with females releasing 1–2 million eggs per spawn. Larvae have a planktonic veliger stage.

Hiatellidae are infaunal suspension feeders. Smaller species nestle into crevices between rocks or bore into soft rocks. Larger species burrow deeply into sand, at the rate of about 12 in (300 mm) per year, until they reach a depth of 3 ft (0.9 m). The oldest recorded specimen of *Panopea generosa* was 168 years old. The family Hiatellidae has a fossil record extending to the Permian.

ADAPEDONTA—SOLENOIDEA—PHARIDAE
RAZOR SHELLS

The family Pharidae, which first appeared during the Cretaceous, contains bivalves that are highly specialized for rapid burrowing into shallow-water sand and mud bottoms, where they live in vertical burrows (which may be two or more times as long as the shell).

Shells of most species are thin, fragile, and greatly elongated along the anterior-posterior axis, rectangular to ovate in outline, and cylindrical to laterally compressed. Valves are identical in size and shape and may gape at both ends. The umbone and small hinge are near anterior end. The shell is aragonitic, and the interior is not nacreous. The pallial line may have a short to moderately deep pallial sinus.

The posterior adductor muscle is smaller than the anterior adductor muscle. Mantle margins are fused ventrally with a large anterior pedal gape. Inhalant and exhalant siphons are long, and may be separate or fused, depending on the genus. Gills are eulamellibranch, and occupy the posterior third of the mantle cavity. The foot is very long, narrow, laterally compressed, and has a terminal bulb. They are capable of very rapid burrowing by inserting

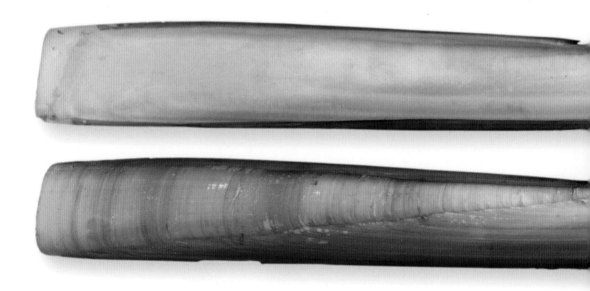

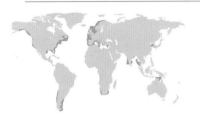

DISTRIBUTION
Global, along temperate and tropical shores, at depths ranging from the intertidal zone to the outer continental shelf (less than 650 ft / 200 m). The genus *Pharella* has been reported from brackish and nearshore freshwater habitats.

DIVERSITY
Family includes 65 living species assigned to 13 genera within 3 subfamilies.

HABITAT
Burrowers, creating vertical or oblique burrows several times the length of their shell, and are capable of moving rapidly up and down the burrows.

SIZE
Species range in size from 2–10 in (50–250 mm).

DIET
Filter feeders.

RIGHT | *Ensis directus* burrowing into the sand on a beach in Belgium.

BELOW | *Ensis siliqua*, a species that inhabits the northeastern Atlantic and Mediterranean Sea, is the largest species in the family Pharidae.

their foot into the sand, then inflating the foot's terminal bulb, expelling the fluidized sand above it. Some may flee their burrows to escape predators, swim a short distance and burrow very rapidly. When placed on the surface, animals can usually be completely buried in a minute or less.

REPRODUCTION
Sexes are separate. Broadcast spawning. Larvae remain in the plankton for several weeks before undergoing metamorphosis and settling to the ocean bottom.

CARDIDA—CARDIOIDEA—CARDIIDAE
HEART COCKLES, GIANT CLAMS

The family Cardiidae is an extremely wide-ranging and diverse lineage. Most species inhabit shallow tropical and temperate seas around the world, burrowing just below the surface of sand or mud bottoms, from intertidal to bathyal depths (0–1,150 ft / 0–350 m). Some live in brackish estuaries and several have adapted to relatively freshwater habitats in the Black Sea and Caspian Sea basins.

BELOW | *Corculum cardissa* lives on sandy bottoms near coral reefs in the tropical Indo-Pacific. The shell has many translucent areas that allow light to support symbiotic algae growing in its gills—this provides nutrition to the animal.

Cardiidae have a heart-shaped outline when viewed from the anterior end of the shell and are often referred to as heart cockles. Giant clams, which had been classified as a separate family, have recently been included within the Cardiidae based on molecular studies.

The oldest fossil identified as a cardiid is from the Triassic.

Shells are equivalve and inflated, solid to translucent, radially sculptured to varying degrees, often with spines or scales on the ribs. Shell margins are serrated, the degree depending on the prominence of the ribs, and serve to align and interlock the valves. Some shells are white; others are brightly colored or patterned. The periostracum ranges from smooth to hirsute. Heart cockles tend to be rounded to trigonal in outline, most elongated dorsoventrally. Adductor

DISTRIBUTION
Cosmopolitan, with the majority of genera inhabiting tropical and subtropical latitudes worldwide. Giant clams are limited to the tropics of the Indo-Pacific.

DIVERSITY
Family includes approximately 270 living species assigned to 48 genera within 8 subfamilies.

HABITAT
Most species are shallow burrowers with the posterior end of the shell just below the surface of sand or mud substrates, often in seagrass beds in shallow water. Some, such as *Corculum* and giant clams, are epifaunal.

SIZE
Heart cockle shells are less than 1 in (25 mm) to over 6 in (150 mm). Giant clams (the largest living bivalves) range from 6 in (150 mm) to 5 ft (1.5 m) in length and can weigh over 725 lbs (330 kg).

RIGHT | *Tridacna gigas*, the largest and heaviest species of bivalve, on a shallow sandy bottom in Cenderawasih Bay, West Papua, Indonesia.

BELOW | A specimen of *Dinocardium robustum* from western Florida exposed to show its strong, muscular foot and extended incurrent and excurrent siphons.

DIET
Most are filter-feeding bivalves. Giant clams are filter feeders as juveniles, but adults have symbiotic algae growing within their tissues to provide nutrition.

REPRODUCTION
Most species are either protandric or simultaneous hermaphrodites with external fertilization and planktonic veliger larvae. Giant clams may produce 100 million eggs in one season and can live to be over 100 years old.

muscles are nearly equal in size and the pallial line is simple, without a sinus. In contrast, giant clams are elongated anteroposteriorly and have only a posterior adductor muscle as adults.

Heart cockles have a strong and muscular foot capable of rapid burrowing, vaulting, and even swimming for short distances to escape predators. Inhalant and exhalant siphons are surrounded by sensory papillae, tentacles, and, in some species, eyes. Giant clams are sedentary and byssally attached to reefs, gaping to expose tissues to sunlight in order to grow symbiotic algae.

CARDIIDA—TELLINOIDEA—TELLINIDAE
TELLINS

Tellinid shells are elongated oval to quadrangular, thin-walled, laterally compressed, slightly inequivalve, with the posterior end of both valves slightly twisted to the right. The hinge plate is small. External sculpture is usually smooth, with fine commarginal growth lines. The exterior of some species may be brightly colored, some with radial rays of color. The periostracum is thin and varnish-like. Anterior and posterior adductor muscle scars are similar in size, and the pallial line is pronounced, with a deep pallial sinus.

The mantle margins are not fused ventrally. Incurrent and excurrent siphons are very long, separate, and move independently. A pair of cruciform (X-shaped) muscles, which is a characteristic of Tellinidae, is present at the base of the incurrent siphon. The gills are eulamellibranch. The foot may be small or large, laterally compressed, and wedge-shaped.

Tellinidae live in deep burrows several times their shell length in soft sediments in marine and estuarine habitats. Some species are oriented vertically, with anterior end down. Others are oriented horizontally, with the right valve upward, which results in the posterior end of the shell being slightly twisted by the siphons. Tellinidae are primarily suspension feeders, but are capable of deposit feeding, using the long siphons.

The superfamily Tellinoidea, which dates to the Triassic, is the most species-rich and diverse lineage in Bivalvia. Earliest records of Tellinidae, one of five families within Tellinoidea, are from the Upper Cretaceous.

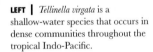

LEFT | *Tellinella virgata* is a shallow-water species that occurs in dense communities throughout the tropical Indo-Pacific.

DISTRIBUTION
Global, with greatest diversity in the tropics.

DIVERSITY
Family includes approximately 500 living species assigned to 107 genera within 9 subfamilies.

HABITAT
Deep burrowers into sand and mud bottoms, extending their long siphons to the surface. They inhabit coastal waters from the intertidal zone to depths of 650 ft (200 m).

SIZE
Species range in size from ¼–5 in (5–127 mm).

DIET
Primarily suspension feeders but may also feed on organic material in surface deposits using their inhalant siphon.

ABOVE | *Macomangulus tenuis* is a common species along the coast of northwestern Europe and the Mediterranean Sea, where it burrows in sandy bottoms.

RIGHT | Shell of *Tellinella cumingii*, a subtidal species from the tropical western Atlantic that orients its shell horizontally within its burrow.

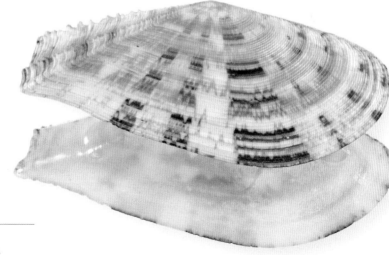

REPRODUCTION
Sexes are separate in most species, although protandric hermaphroditism has been reported in some taxa. Some produce benthic jelly-encased spawn that develop into planktonic veliger larvae.

CARDIIDA—TELLINOIDEA—DONACIDAE
WEDGE SHELLS

Most members of the family Donacidae are small, active bivalves adapted for life on open sandy beaches of tropical and subtropical regions. Many donacids occur in very large populations. As the waves wash the animals onto the beach, they re-burrow rapidly, and migrate shoreward on incoming tides and seaward with receding tides, filtering food from the nutrient-rich turbulent waters. These species may co-occur with a second species of donacid that is smaller, lives slightly offshore, and does not migrate with the tides.

Shells are strong, equivalve, triangular and wedge-shaped, with the anterior end longer, narrower, and smoother, and the posterior end obliquely truncated and broader, with more pronounced sculpture, primarily of radial threads or cords. The heterodont hinge is near the posterior margin of the shell. Anterior and posterior adductor muscles scars are nearly equal in size, the pallial line is strong, with a distinct pallial sinus. Shells of many species are brightly colored, both externally and internally, often with radial rays of alternating colors on the exterior that vary in color and pattern among animals within a population. It has been hypothesized that such variation may be an adaptation to prevent birds, the main predators of donacids, from forming a single search image.

Many species of Donacidae are harvested for food and bait throughout the

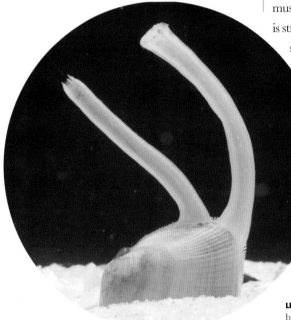

LEFT | *Donax variabilis* partially buried in sand with siphons extended. Sanibel Island, Florida.

DISTRIBUTION
Sandy beaches at tropical and temperate latitudes worldwide. Estuarine and freshwater genera are limited to South America and Africa.

DIVERSITY
Family includes 109 living species assigned to 5 genera within 1 subfamily.

HABITAT
Many species live in the surf zone on sandy beaches, moving onshore and offshore with the changing tides. Others live in slightly deeper, subtidal sand bottoms. Nearly a dozen species are limited to brackish or freshwater environments.

SIZE
Species range in size from ½–2 in (12–50 mm), but some may reach 6 in (150 mm).

world. The species *Donax variabilis*, commonly known as the Coquina, is abundant in the southeastern United States, where it is made into a stew.

The family Donacidae has a fossil record that extends to the Early Cretaceous (Aaptian 121–113 mya).

TOP | *Donax variabilis* exposed by waves along sandy beach in Ocracoke, North Carolina.

RIGHT | *Latona madagascariensis* is a species with distinctive surface sculpture that inhabits shorelines along southern and eastern Africa and Madagascar.

BELOW | *Donax culter* is one of several species along the shores of the tropical eastern Pacific from the Gulf of California to Panama.

DIET
Suspension feeders.

REPRODUCTION
Sexes are separate. Fertilization is external. The pelagic veliger larvae may spend several weeks to 2 months in the plankton.

MYIDA — PHOLADOIDEA — PHOLADIDAE
PIDDOCK CLAMS

BELOW | The shell of *Cyrtopleura costata*, a common Florida species, showing the mesoplax.

Members of the family Pholadidae are infaunal bivalves that burrow to a depth that exceeds their shell length. They have long fused siphons (often several times the length of the shell) that extend above the burrow surface and cannot be retracted into the shell. Shells may be thin or thick, depending on the substrate into which it burrows. Shells are symmetrical, generally elongated, and gape widely at both ends, although some taxa seal their anterior gape with two calcareous plates (callum) upon reaching adult size. Most have ribs or scales that may be limited to the posterior portion of the shell and assist with burrowing into the substrate. Shells have strong anterior and posterior muscle scars and lack a hinge, with few teeth or ligaments.

Some pholadids have accessory shell plates such as the protoplax, which covers the anterodorsal margin of the shell. The mesoplax, present in most pholadids, is situated above the beaks of the two valves. Most genera also have apophyses, spoon-shaped structures below the umbos,

DISTRIBUTION
Most species inhabit shallow waters along shorelines and range from cold to temperate and tropical regions. Some species burrow in sunken wood and have been reported from depths of over 23,000 ft (7,000 m).

DIVERSITY
Family includes approximately 140 species assigned to 16 genera within 3 subfamilies.

HABITAT
Burrowers in sand, mud, clay, shale, or coral, mostly in intertidal to shallow subtidal depths. Often occur in dense populations.

SIZE
Species range in size from 1–8 in (25–200 mm), but the larger animals cannot completely retract into their shells.

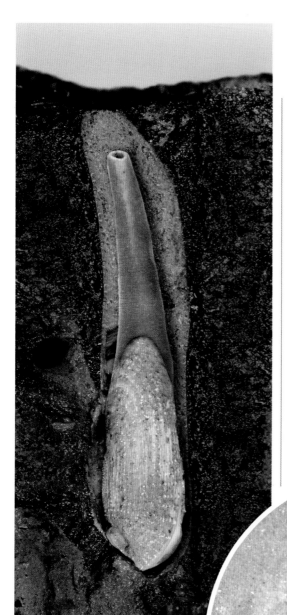

to which the foot muscles are attached. There may also be an extension of the posterior margin of the shell that covers the siphons. Gills are large and extend into the long siphons that lack pallets at the siphon tips.

Fossil burrows attributed to Pholadidae are known from the Carboniferous (359–299 mya). Individuals of some of the larger species have been reported to live between five and eight years.

LEFT | *Barnea candida* is a European species that burrows in peat and clay.

BELOW | *Cyrtopleura costata*, exposed siphons. Kice Island, Collier County, Florida.

DIET
Suspension feeders that filter food through very large gills, which extend into the very large siphon.

REPRODUCTION
The majority are protandric hermaphrodites (initially males, becoming females). Fertilization may be external via broadcast spawning or internal, with some species brooding their larvae.

MYIDA—PHOLADOIDEA—TEREDINIDAE
SHIPWORMS

Shipworms settle on wood as larvae, burrow into it by mechanical abrasion using the surface of their small shells, and burrow deeper as they grow. They have a long, worm-like body, with a greatly reduced shell at the anterior margin of animal at the base of the burrow. Vital organs are situated posterior to the shell. The gill consists only of inner demibranchs. The burrow has a calcareous lining and increases in size as the animal grows.

Calcareous paddle-shaped structures called pallets are present at the tips of the siphonal and close the burrow when the siphons are retracted. Teredinids consume the wood through which they burrow and symbiotic bacteria living within their gills produce enzymes that convert the cellulose of wood into sugars that provide its nutrition. They may also filter feed to supplement their diet.

Some species burrow in sediment rather than wood. *Kuphus polythalamius* lives in deep burrows among mangroves and has been reported to feed on particles filtered from the water. More

BELOW | The shell of *Teredo navalis*, situated at the anterior end of the animal.

DISTRIBUTION
Global, primarily in shallow, temperate, and tropical coastal waters. Some species occur on sunken wood at depths of over 2,300 ft (700 m).

DIVERSITY
Family includes approximately 100 living species assigned to 18 genera within 3 subfamilies.

HABITAT
Most species burrow in wood; some in soft sand and mud bottoms.

SIZE
Shells range in size from 1/16–2 in (3–50 mm), but tubes are much larger, ranging from 1 in (25 mm) to 5 ft (1.5 m) in length and 4 in (100 mm) in diameter.

ABOVE | *Teredo navalis* animals in wood.

RIGHT | Anterior end of *Nivanteredo coronata* showing position of shell and foot.

recent reports indicate intracellular symbiotic sulfur-oxidizing bacteria within its gills contribute to its nutrition.

The family Teredinidae have been reported from Cretaceous fossils. Shipworms have long been known for the damage they cause: they had infested and damaged wooden ships of the Greek and Roman navies and caused Columbus to abandon two ships on one of his voyages to the Caribbean. Wooden ships likely contributed to the broad geographic distribution of some species.

DIET
Most species rely on symbiotic bacteria to produce enzymes that convert the cellulose of the wood that they consume into sugars.

REPRODUCTION
Most species are protandric hermaphrodites (initially males, becoming females), although some are simultaneous hermaphrodites (with the same animal producing both eggs and sperm simultaneously). Fertilization may be external or internal.

VENERIDA—VENEROIDEA—VENERIDAE
VENUS CLAMS

With origins in the Cretaceous, the Veneridae is the most species-rich and diverse family among the Bivalvia, with more than 750 living species, although ongoing studies continue to refine their taxonomy and the phylogenetic relationships among them. A large majority of Venus Clams are infaunal, active, and rapid burrowers in sand and mud bottoms, primarily at intertidal and shallow depths along tropical and subtropical shores throughout the world. Some small species can occur at densities approaching 100,000 individuals per square yard or meter of mud bottom.

Shells occur in a variety of shapes. Most are circular, ovate, or trigonal in outline, equivalve, and may be inflated or compressed, thick or thin with a surface that ranges from smooth to sculptured with radial ribs, concentric bands, or a cancellate surface. A few have extensive spines on the posterior portion

RIGHT | *Bassina disjecta*, commonly called the Wedding Cake Venus, burrows in shallow-water sandy bottoms off the southeastern coast of Australia.

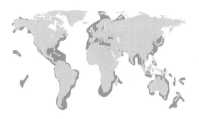

DISTRIBUTION
Global, most in temperate to tropical latitudes but some extend to polar regions and bathyal depths.

DIVERSITY
Family includes approximately 750 living species assigned to 107 genera within 13 subfamilies.

HABITAT
Most species are infaunal burrowers in mud and sand in marine and estuarine environments. Some are attached by a byssus to substrates, either infaunal or epifaunal; others burrow within clay, coral, or soft rocks.

SIZE
Species range in size from $1/16$–7 in (2–170 mm).

LEFT | *Polititapes rhomboides* using its muscular ventral foot along a shallow-water bottom off Galicia, Spain.

BELOW | *Lioconcha castrensis* is a common species in shallow, sandy bottoms throughout the tropical Indo-Pacific, and is harvested for food in some areas.

of the shell. All have a heterodont hinge with three cardinal teeth, an external posterior ligament, and anteriorly directed umbones. Anterior and posterior adductor muscles are roughly equal in size and the pallial sinus may be shallow or deep. The periostracum may be thin or thick. A byssus is usually absent in adults.

The animals have a large foot capable of rapid burrowing. Siphons are present, may be separate or united, and are usually of short to moderate length. Some species have life spans as long as forty years. Many venerid species are edible and are commercially fished throughout the world and some are maricultured. Native Americans had used the shell of the species *Mercenaria mercenaria* to produce beads used as currency, called wampum.

DIET
Filter feeders.

REPRODUCTION
Sexes are separate in most venerids. Fertilization is external, with planktonic larvae. Some species are protandric hermaphrodites. Several species brood their larvae within their gills; others deposit eggs in gelatinous capsules that hatch as direct developing juveniles.

VENERIDA—ARCTICOIDEA—ARCTICIDAE
OCEAN QUAHOGS

The family Arcticidae arose during the Triassic (252–201 mya), diversified during the Cretaceous, and had a global distribution, but only a single genus—*Arctica*—with a single species—*Arctica islandica*—survives in the living fauna, confined to boreal waters along both coasts of the North Atlantic Ocean. It is a commercially harvested species.

The shell is large, ovate, and equivalve, with a smooth surface covered by a thick brown or black periostracum. The mantle is open anteriorly, the pallial sinus is absent, the thick foot has a ventral groove, and gill lamellae are broad. This species lives in dense beds on level bottoms of medium- to fine-grained sand at depths limited by temperature (below the 16°C /60° F isotherm) during the summer months. Animals are suspension feeders filtering phytoplankton. Growth is rapid during the first two to three years of life but slows, and adults of similar size may vary greatly in age. Reproductive maturity is reached between seven and thirteen years of age. A specimen dredged off Iceland in 2006 was determined to be 507 years old. It is the oldest individual (non-colonial) animal known. It was informally named Ming the Clam because it started its life during the Ming Dynasty in China.

DISTRIBUTION
North American and European coasts of the North Atlantic at depths of 45–840 ft (15–255 m).

DIVERSITY
Family includes 1 living species assigned to the genus *Arctica*.

HABITAT
Fine to coarse sandy bottoms.

SIZE
Species range in size from 2½–5 in (64–127 mm).

DIET
Filter feeders.

REPRODUCTION
Sexes are separate. Fertilization is external, with spawning occurring from spring to fall. Larvae have a planktotrophic stage prior to metamorphosing and settling to the ocean bottom and burrowing.

ABOVE | *Arctica islandica*. An exposed animal laying on its right side, showing its siphons and surrounding tentacles.

RIGHT | *Arctica islandica*. The shell is burried in the sand, with only the siphons exposed.

OPPOSITE | Shell of *Arctica islandica*. Dredged on a sandy bottom off the coast of New Jersey, at a depth of 195 ft (65 m).

VENERIDA—CHAMOIDEA—CHAMIDAE
JEWEL BOX CLAMS

Members of the family Chamidae have adapted to an epifaunal habitat, with their shells attached to subtidal hard substrates. Species of the genus *Chama* have been recognized on the basis of their being cemented throughout their life by their left valve, while species of *Pseudochama* are

BELOW | *Chama* sp. with siphons extended off Tulamben, Bali, Indonesia.

DISTRIBUTION
Mostly tropical and temperate seas, living on coral or rocky shores in shallow water.

DIVERSITY
Family includes 66 living species assigned to 6 genera within 1 subfamily.

HABITAT
Most species are sessile and epifaunal, inhabiting clear marine waters in the sublittoral zone.

SIZE
Species range in size from 1½–3 in (38–75 mm); some with long spines may reach 5 in (125 mm).

cemented by their right valve. However, more recent studies indicate that this can vary. Species of *Arcinella* are attached as juveniles, but detach and become free-living as adults, laying on the bottom with either side uppermost.

The shells are solid, subcircular, and irregular in outline. The valve that is attached is deeply concave while the valve that is uppermost tends to be flatter. Species of *Arcinella* are more nearly equivalve with both valves rounded. The shell exterior and especially the uppermost valve may be prominently sculptured with radial ribs, spines, and commarginal frills. The shell interior has both anterior and posterior adductor muscle scars as well as a pallial line. The foot is small, laterally flattened, and extensible, and is used to clean the eulamellibranchiate gills and mantle cavity.

Chamidae inhabit nearshore waters in temperate and tropical regions, along rocky shores and coral reefs from subtidal depths to depths of about 100 ft (30 m). They are suspension feeders, but do not tolerate turbid water or low salinity.

The superfamily Chamoidea consists of only a single family, Chamidae. Earliest records are from the Upper Cretaceous.

ABOVE | *Arcinella cornuta* from Sanibel Island, Florida.

DIET
Filter feeders.

REPRODUCTION
Sexes are separate, but some species may alternate between male and female. Most species have external fertilization with small eggs; larvae have a long planktonic phase before settling.

VENERIDA—GLOSSOIDEA—VESICOMYIDAE
VESICOMYAS

Bivalves of the family Vesicomyidae have become adapted for life in sulfide-rich reducing environments that occur at hydrothermal vents along ocean spreading ridges and converging plate boundaries, as well as at hydrocarbon seeps and whale falls. Like many other members of these "extremophile" communities, which survive extreme temperatures, high pressure, perpetual darkness, and toxic chemicals dissolved in the water, vesicomyids have evolved a symbiotic relationship with sulfide-oxidizing bacteria that live within their gills and provide all or part of their nutrition.

Species of Vesicomyidae are widely distributed throughout the world from the outer continental shelf to hadal depths. The subfamily Pliocardiinae is the most diverse and contains fourteen of the fifteen genera that include living species. Most of the species have large shells, some reaching 12 in (300 mm) in length, which are oval to kidney-shaped, laterally compressed, with an anteroventral pedal gape and a chalky exterior with thick periostracum that is usually worn. The mantle and foot are pink, and the visceral mass is red due to hemoglobin in the blood. Siphons are short, gills are large and thick with cells that contain symbiotic chemoautotrophic

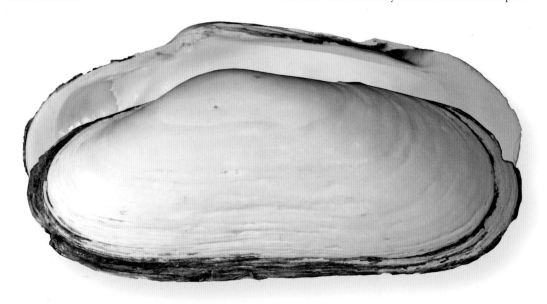

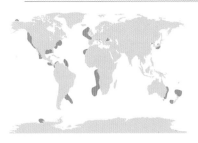

DISTRIBUTION
Global, from outer continental shelf to hadal depths (330–29,500 ft / 100–9,000 m).

DIVERSITY
Family includes approximately 150 living species assigned to 15 genera within 2 subfamilies.

HABITAT
Along hydrothermal vents and hydrocarbon seeps. Adapted to sulfide-rich reducing habits.

SIZE
Species range in size from ½–10 in (10–250 mm) or more.

DIET
Nutrients produced by chemoauthotropic bacteria living in the gills of vent species.

REPRODUCTION
Sexes are separate. Larvae develop from large, yolky eggs and are dispersed over large distances by bottom currents.

OPPOSITE | *Calyptogena magnifica* from hydrothermal vents in 6,560 ft (2,000 m), along the Galapagos Rift, Galapagos Islands, Ecuador.

ABOVE | A colony of *Calyptogena magnifica* near hydrothermal vents at abyssal depths along the east Pacific Rise and Galapagos Rift.

bacteria. Animals absorb nutrients produced by bacteria and have guts that are reduced or absent.

The subfamily Vesicomyinae includes a single genus *Vesicomya* that is presently known from about twenty living species that are usually small (less than ½ in / 10 mm), have a gut, lack specialized gill cells (subfilamental tissues) containing bacteria, and are not restricted to sulfide-rich habitats. The family Vesicomyidae has been a dominant component of chemosynthetic ecosystems since the late Eocene.

VENERIDA—MACTROIDEA—MACTRIDAE
SURF CLAMS

Mactridae inhabit sandy shorelines along tropical and temperate continental margins around the world, often occurring in large populations at subtidal depths. One common species, *Spisula solidissima*, is among the largest bivalves along the western Atlantic, and can grow to 11 in (228 mm) in length and reach an age of thirty years. Some species can tolerate reduced salinities and inhabit estuaries. *Rangia cuneata*, a species native to the Gulf of Mexico, is considered to be an invasive species that has appeared in European waters (Belgium, Poland) since 2005.

Shells of most mactrids are trigonal, oval, or elongated in shape, equivalved, and equilateral with an anteriorly deflected umbone. A large, triangular, spoon-shaped internal ligament (resilium) below the umbone is diagnostic of Mactridae. The hinge is heterodont. The shell exterior tends to be smooth, with fine, commarginal growth lines or more pronounced undulations. Radial ridges may occur in some species.

Shell color ranges from uniform white to brown. Radial rays differing in color may be present. The peristracum is thin and usually at least partially abraded in larger specimens. The valve interiors are non-nacreous, with the anterior adductor muscle scar equal in size to the posterior adductor muscle

DISTRIBUTION
Global, primarily in tropical and temperate coastal waters at continental shelf depths of 0–650 ft (0–200 m).

DIVERSITY
Family includes approximately 200 living species assigned to 37 genera within 4 subfamilies.

HABITAT
Infaunal burrowers in sand and mud. Primarily marine and estuarine. Some genera can live in nearly fresh water.

SIZE
Species range in size from 1–11 in (25–275 mm).

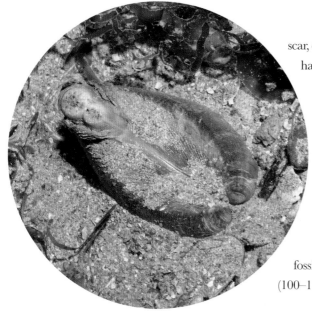

scar, or slightly more elongated. The pallial line has a deep sinus.

There are two retractable siphons that are fused and covered with a sheath of periostracum. Mactrids are capable of rapid burrowing using their muscular, wedge-shaped foot together with water currents produced by contracting the shells. The gills are eulamellibranch with the outer demibranchs reduced in size.

The family Mactridae first appeared in fossils dating to the Lower Cretaceous (Albian) (100–113 mya).

ABOVE | *Mactromeris polynyma* partially buried in a seagrass bed in shallow water off Singapore.

LEFT AND OPPOSITE | *Raeta plicatella* is a common intertidal species along the shores of the western Atlantic from New Jersey to Argentina. It has a thin shell reinforced by the undulating corrugated surface.

BELOW | *Mactra violacea* is a large and abundant shallow-water species harvested commercially in the Indian Ocean.

DIET
Filter feeders.

REPRODUCTION
Sexes are separate. In some species, fertilization and early stages of larval development occur within the mantle cavity of females before pelagic larvae are released.

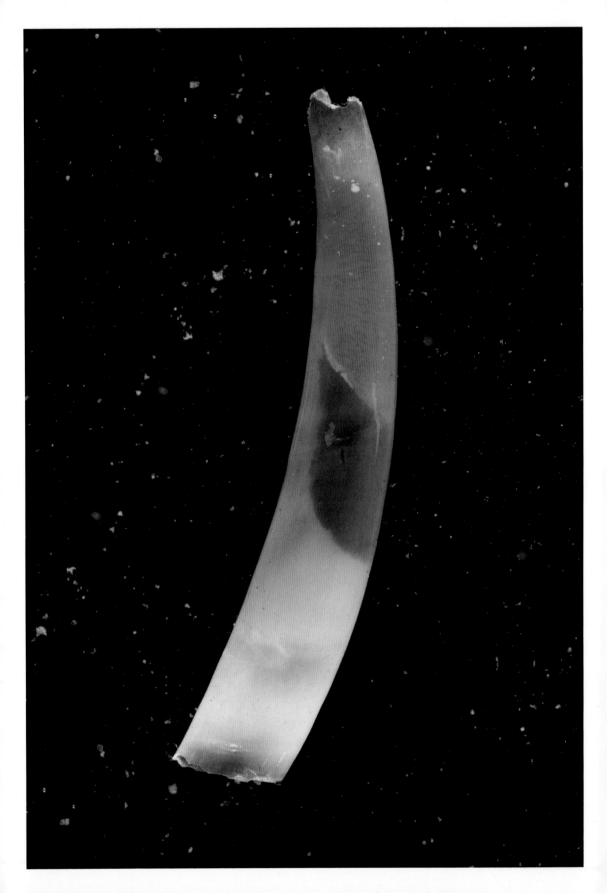

OPPOSITE | A scaphopod mollusk collected at a depth of 1500 ft (500 m) in the Bellingshausen Sea, off Antarctica.

SCAPHOPODA

Scaphopods, also known as tusk shells due to their resemblance to elephant tusks, are bilaterally symmetrical, infaunal micropredators with a curved, tubular shell, open at both ends.

The larger shell opening is ventral; the smaller opening, which may have a notch, is dorsal. The anterior part of the shell is concave, the posterior is convex. Shells are $1/16$–6 in (2–150 mm) in length. The shell is composed of aragonite in most taxa. The body is attached to the shell along the concave anterior region and surrounded by the mantle cavity, which is open at both ends. The dorsal end of the mantle forms a sleeve of tissue called the pavillon, which opens along the posterior margin. A valve separates the pavillon from the mantle cavity and regulates the flow of water.

The animals lack gills, and respiration is through the surface of the mantle. The foot extends through the ventral opening. The head has a large snout and lacks eyes. The buccal apparatus includes a radula that is composed of five teeth per transverse row. The teeth may be mineralized with calcium and iron. Scaphopods have multiple thread-like tentacles (captacula) that emerge from the sides of the snout

SHELL FEATURES AND ANATOMY OF A TYPICAL SCAPHOPOD

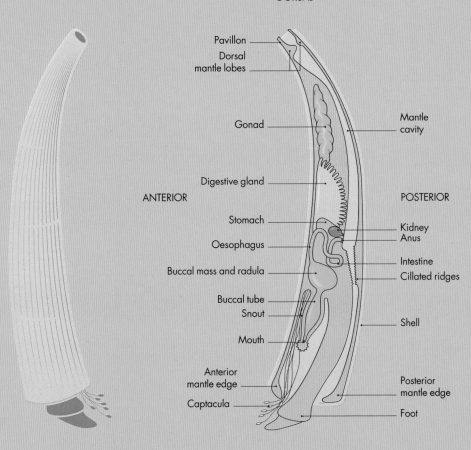

to probe the sediment for food items, such as foraminifera, diatoms, ostracods, and other small invertebrates, then retract to bring them to the mouth.

With origins in the Middle Ordovician, Scaphopoda is believed to be the youngest class of Mollusca. Scaphopods were thought to be the sister group to Bivalvia, or questionably to the extinct class Rostroconchia. More recent studies suggest closer affinities to Gastropoda and Cephalopoda. During the late Paleozoic or early Mesozoic, Scaphopoda diverged into two orders: the Dentaliida, which includes the oldest known scaphopod from the Ordovician, and the younger Gadilida.

Scaphopods inhabit sand and mud bottoms at all latitudes, from the intertidal zone to abyssal depths, with diversity greatest in tropical latitudes and offshore depths. They burrow into the substrate with the ventral end down and with only the tip of the dorsal end protruding above the surface of sediment. As the animal grows by addition of shell to the ventral margin, the dorsal end is periodically broken or resorbed in order to increase the diameter of the dorsal aperture. Populations may have a patchy distribution, yet can be locally abundant with densities of forty or more individuals per square yard.

Scaphopod shells have been used as currency and to decorate clothing by Pacific Northwest Native Americans and First Nations for thousands of years.

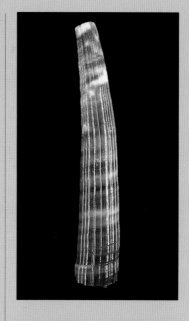

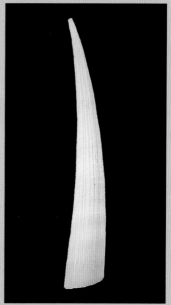

TOP | *Pictodentalium formosum*, a rare shallow-water species from eastern Asia.

ABOVE | *Coccodentalium carduus*, a deep-water (600–2100 ft / 138–640 m) species from off southern Florida and the West Indies.

DENTALIIDA—DENTALIIDAE
TUSK SHELLS

BELOW | From left to right: *Dentalium elephantinum*, *Antalis vulgaris*, and *Graptacme marchadi*.

The class Scaphopoda is divided into two orders: the Dentaliida and the Gadilida. The Dentaliida includes the oldest known scaphopods and has been dated to the Middle Ordovician (470 mya) while the first Gadilida is recorded from the Permian (427 mya).

The Dentaliida includes eight families that are all represented in the living fauna and that include 22 genera and nearly 300 living species. All species of Dentaliida have a characteristic curved, tapering, tubular shell that is open at both ends, with the widest part of the shell being at the anterior aperture. The surface may be smooth or longitudinally ribbed. All have a tapering foot with a pair of extendable lateral lobes that retracts by bending into the shell. The valve separating the pavillon from the mantle cavity is oriented horizontally. The central tooth of the radula is broader than it is high.

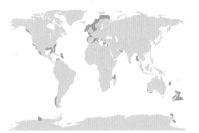

DISTRIBUTION
Most diverse in tropical and subtropical latitudes from subtidal to bathyal depths. Diversity decreases at higher latitudes and greater depths.

DIVERSITY
Family includes approximately 250 living species assigned to 13 genera within 1 subfamily.

HABITAT
Infaunal burrowers in sand and mud bottoms.

SIZE
Living species range in size from ¾–6 in (18–150 mm). Some fossil species reached nearly 12 in (300 mm) in size.

DIET
Interstitial organisms such as foraminifera, diatoms, ostracods, as well as detritus. Some species preferentially feed on foraminifera.

ABOVE | *Pictodentalium festivum*—exposed anterior of shell with foot and lateral lobes extended.

Of the eight families of Dentaliida, Dentaliidae is the oldest and largest, both in terms of diversity and shell size. Longitudinal sculpture varies among genera, with some having strong cords and others fine striae. The posterior opening of the shell varies among genera, with some having a simple longitudinal notch along the anterior side, while others have an elongated slit, a series of perforations, or a calcareous plug.

Dentalliidae is the only family in which the shell color may be other than white, with green and yellow or pink shells occurring in some species. Diversity is greatest in tropical and temperate latitudes, at depths ranging from subtidal to 3,280 ft (1,000 m), with peak diversity at depths of 1,640–2,625 ft (500–800 m).

REPRODUCTION
Sexes are separate. Fertilization is external. Larvae are lecithotrophic, feeding on large yolks before metamphosis and settlement to the ocean bottom.

GADILIDA—GADILIDAE
TOOTH SHELLS

BELOW | *Gadila* sp. animal visible by transparency.

The order Gadilida is younger than Dentaliida, with earliest records attributed to the Permian, but with most of the four living families first appearing in fossil deposits ranging from the Triassic to the Cenozoic. Shells of Gadilida are distinguished from those of Dentaliida in being smaller, with the widest part of the curved tubular shell being behind the aperture, in some cases near the middle of the shell. The surface is smooth and glassy, lacking strong longitudinal sculpture. The foot also differs in having a circular terminal disk with papillae along its circumference and a terminal papilla at its center. The foot can extend to twice the length of the shell and withdraw into the shell by introversion. The valve separating the pavillon from the mantle cavity is oriented vertically. The central tooth of the radula is taller than broad and may have a cusp.

The family Gadilidae is the largest of the four families of Gadilida. The shells tend to be small (¼–1 in / 6–25 mm), only slightly curved, circular, or slightly compressed in cross-section, and may have two to four slits along the dorsal aperture. They tend to be smooth, glassy, and in some cases translucent, with the animal visible through the shell. Gadilidae are microcarnivores that feed on a variety of small organisms including foraminiferans and

DISTRIBUTION
Global, ranging from subtidal to abyssal depths.

DIVERSITY
Family includes approximately 200 living species assigned to 9 genera within 1 subfamily.

HABITAT
Burrowers in sand and mud bottoms.

SIZE
Species range in size from ¼–2 in (6–25 mm), but most species are less than 1 in (25 mm) in length.

DIET
Foraminifera, mites, and larval invertebrates in the sediments.

diatoms as well as other small organisms that they capture using their extensible captacula.

As with Dentaliidae, the diversity of Gadilidae is greatest in the tropics but extends to polar regions at depths from subtidal to abyssal. However, highest diversities occur at greater depths (3,900–6560 ft / 1,200–2,000 m).

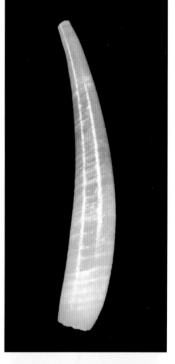

RIGHT | *Polyschides pelamidae*, a species from the tropical western Pacific at depths of 30–150 m (90–450 ft).

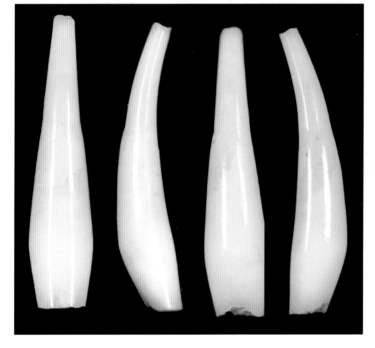

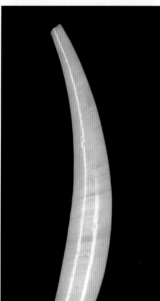

REPRODUCTION
Sexes are separate. Fertilization is external. Larvae are lecithotrophic and pelagic before settling to the ocean bottom to metamorphose after 3–5 days. Larval shells are dissolved or broken during subsequent growth.

ABOVE | Multiple views of a specimen of *Gadila watsoni*, dredged in the Yucatán Strait at a depth of 1170 m (3830 ft).

RIGHT | *Gadila zonata*, a Philippine species from gray mud bottoms at depths of 1500–3000 ft (500–1000 m)

OPPOSITE | *Naticarius stercusmuscarum*, a common shallow-water species throughout the Mediterranean.

GASTROPODA

With about 90,000 living species, Gastropoda comprise the largest class of Mollusca and one of the most diverse groups of animals living today, second only to the insects (class Insecta). Gastropods are also the most ecologically diverse mollusks, occurring in all marine habitats, and have given rise to multiple independent radiations into freshwater and terrestrial environments.

Gastropoda differ from all other molluscan classes in having undergone torsion, a process that occurs in the larval stage and results in the body being twisted up to 180° in a counterclockwise direction. Torsion reorients the formerly posteriorly directed mantle cavity and the organs it encloses, including paired gills, kidney, and reproductive openings as well as the anus, to a position above and behind the head of the animal. This allows the animal to retract its head into its shell and improves water flow through the mantle cavity. As a result, gastropods are not bilaterally symmetrical. However, some members of the more derived lineages within Gastropoda (opisthobranchs and pulmonates) undergo a process of detorsion, in which the mantle cavity and anus revert to a posterior position.

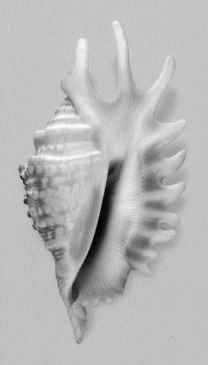

ABOVE | *Ophioglossolambis violacea* inhabits offshore sand bottoms (20–60 ft / 7–20 m) in the Eastern Indian Ocean.

Most gastropods are easily recognized by their single conical shell that may be low and limpet-like or long and helically coiled around an axis in a right-handed spiral. Shells may vary among lineages, some with a tall spire and numerous whorls; others with a very low spire with few, rapidly expanding whorls. Shells range in size from a fraction of a millimeter to 3 ft (1 m) in shell length. Although all gastropods evolved from an ancestor with an external shell, in some lineages the shell has become internal; in others the shell is entirely lost. An operculum is present in nearly all larval gastropods and is retained in some lineages but lost in others. When present in adults, opercula range from being thin, chitinous, and translucent to thick and heavily calcified.

The gastropod body consists of a head with a terminal mouth, tentacles, and eyes; a visceral mass (which includes a gonad, digestive system, heart, and kidney) that is wound within the shell and covered by mantle epithelium that secretes the shell; and a large, muscular foot with a creeping sole. In most gastropods the animal is capable of completely withdrawing into the shell.

Gastropods occur in all marine habitats from the splash zone along shorelines to the deepest trenches. While most are benthic animals, several groups have independently become adapted to a pelagic existence throughout the oceans. Multiple lineages have radiated into freshwater habitats while others developed lungs and colonized terrestrial environments from nearshore shrubbery to forests, deserts, and mountaintops. Different lineages adapted a variety of diets, many feeding on algae and detritus, some becoming parasitic, and others becoming predators of various invertebrates, with some capable of hunting fish.

Reproductive strategies are also very diverse among gastropods. Some have separate sexes; others are hermaphrodites (some consecutive, others simultaneous). Some shed gametes into the waters; others have internal fertilization. Some brood their young; others deposit eggs in elaborate proteinaceous egg capsules, with young that may hatch as planktonic larvae or crawling juveniles.

The first appearance of Gastropoda in the fossil record was in the Lower Cambrian (541 mya), and many lineages and clades have continued to diversify since then. Research into the relationships among these lineages is ongoing, with over 240 extinct and nearly 500 living families currently recognized.

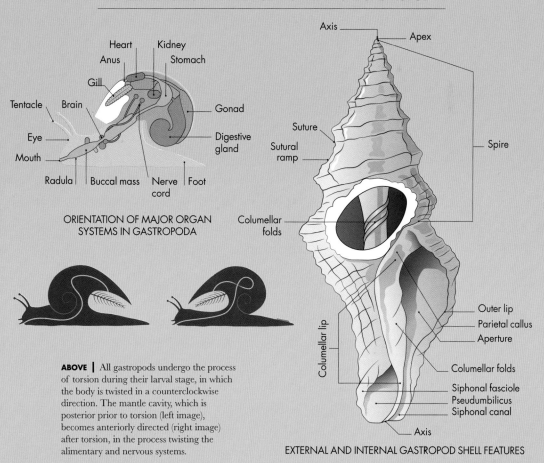

SHELL FEATURES AND ANATOMY OF A TYPICAL GASTROPOD

ORIENTATION OF MAJOR ORGAN SYSTEMS IN GASTROPODA

ABOVE | All gastropods undergo the process of torsion during their larval stage, in which the body is twisted in a counterclockwise direction. The mantle cavity, which is posterior prior to torsion (left image), becomes anteriorly directed (right image) after torsion, in the process twisting the alimentary and nervous systems.

EXTERNAL AND INTERNAL GASTROPOD SHELL FEATURES

PATELLOGASTROPODA—PATELLOIDEA—PATELLIDAE
TRUE LIMPETS

The Patellidae is the most basal family within Patellogastropoda, which is the most primitive group of Gastropoda living today. The patellid shell is conical, not coiled, generally with a central or slightly anterior apex from which radial ribs that vary in prominence originate and may be intersected by rugose commarginal growth lines. It is bilaterally symmetrical and may be tall or low and broad, with an oval, elliptical, or star-shaped outline. The interior is porcelaneous, usually white, often darker beneath the apex, with a U-shaped muscle scar that is interrupted anteriorly. An operculum is present in veliger larvae but absent in adults.

The animal has a large, muscular foot. The head has a snout and a pair of tentacles with eyes at their bases. The mantle is continuous around the perimeter of the shell. Its outer, glandular region secretes the shell, and the outer edge of the mantle has a row of sensory tentacles.

Patellids lack a true gill. The inner region of the mantle is involved in respiration and may be smooth or have a fringe of respiratory tentacles or leaflets. Patellids are sedentary animals and often occur in large populations (some >1,000 per square yard or meter) on intertidal hard substrates. Many excavate a shallow area beneath them to which their shell conforms, producing a tight seal. They forage for

DISTRIBUTION
Primarily antitropical, with most species inhabiting shores along South Africa and the North Atlantic. Other species are more widely distributed throughout temperate or tropical regions.

DIVERSITY
Family includes 47 living species assigned to 4 genera within 1 subfamily.

HABITAT
Most species live on intertidal rocky shores.

SIZE
Species range in size from 1–2½ in (25–60 mm), but *Scutellastra mexicana* can reach 14 in (350 mm) in length.

DIET
Algae and lichens that grow on neighboring rocks.

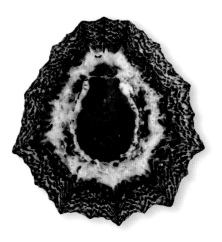  

ABOVE | Left to right: Dorsal views of the shells of *Cymbula granatina*, *Helcion pruinosus*, and *Scutellastra barbara*, all common species on rocky shorelines along the coast of southern Africa.

food, scraping algae and lichens from nearby areas with their radula, portions of which may be mineralized, but return to their home scar after feeding. Some exhibit territorial behavior, driving other limpets and species from their territory.

Compared to other mollusks, patellids have small genomes that span nine pairs of chromosomes. They were the first mollusks to have their genome sequenced. The earliest records of Patellidae date to Permian deposits (260–245 mya).

OPPOSITE | A cluster of *Patella vulgata* on rocks exposed at low tide.

RIGHT | Underside of the limpet *Patella vulgata*, showing foot, snout, cephalic tentacles, and mantle edge.

REPRODUCTION
Sexes are separate in some species; others are protandrous or simultaneous hermaphrodites. Fertilization is external. Eggs hatch as planktonic larvae.

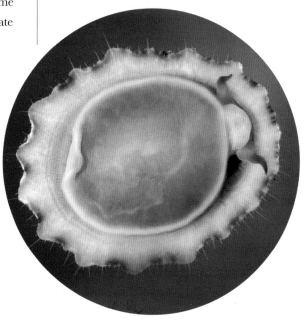

PATELLOGASTROPODA—LOTTOIDEA—LOTTIIDAE
LIMPETS

Lottiidae is the most diverse and abundant of patellogastropods, occurring primarily at all cool temperate to tropical latitudes barring the Antarctic coast. Species and genera are most diverse along high intertidal rocky shores, but also occur in estuaries and in seagrass beds.

Shells are conical, typically with apices that are subcentral or slightly anteriorly displaced, tall or relatively flattened, depending on their habitat, and circular to star-shaped in outline. Exterior sculpture may be smooth, radially ribbed, or with fine commarginal bands. The shell interior is porcellaneous. An operculum is present in larvae, but not in adults. Lotiids may be distinguished from other patellogastropods based on their shell's crystalline composition. They lack a calcitic foliated shell structure present in the other families. Unlike patellids, lotiids have a single left gill that extends over their head in the shallow anterior portion of their mantle cavity. Some of the larger species may also have secondary gills in the regions of the mantle cavity surrounding the foot. Lottiids can be distinguished based on other anatomical differences in various organ systems, and on the basis of the morphology of their radular teeth.

They are primarily herbivores, scraping algae and algal films with their radulae. Some species return to a home scar when the tide recedes. *Lottia gigantea*, a large species that may reach nearly 4 in (100 mm) in length, defends its territory of about 328 ft^2 (100 m^2), on which it maintains a meadow of algae.

The fossil record indicates that the family Lottidae first appeared during the Early Cretaceous (Albian 113–100.5 mya), slightly later than Patellidae, and has undergone significant radiation since the Pliocene.

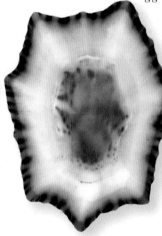

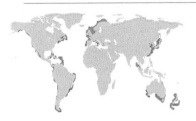

DISTRIBUTION
Shallow marine environments worldwide, except around Antarctica.

DIVERSITY
Family includes approximately 150 living species assigned to 15 genera within 2 subfamilies.

HABITAT
Occupy hard substrates in a wide range of intertidal habitats, including brackish estuaries. Few species occur at depths greater than 100 ft (30 m).

SIZE
Species range in size from ³⁄₁₆–5 in (5–125 mm).

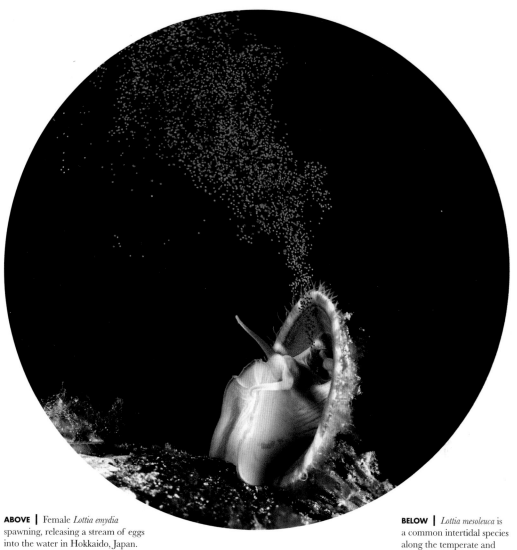

ABOVE | Female *Lottia emydia* spawning, releasing a stream of eggs into the water in Hokkaido, Japan.

OPPOSITE | *Patelloida saccharina* is an abundant limpet inhabiting intertidal and subtidal rocks throughout the tropical Indo-West Pacific.

BELOW | *Lottia mesoleuca* is a common intertidal species along the temperate and tropical eastern Pacific.

DIET
A variety of algae, including algal films.

REPRODUCTION
Protandric hermaphrodites that are males first become females with increasing age. Fertilization is external. Eggs have sufficient yolk to support larvae through pelagic stage, metamorphosis, and settlement as juveniles.

NEOMPHALIDA—NEOMPHALOIDEA—PELTOSPIRIDAE
TAPERSNOUTS

The order Neomphalida contains a single superfamily, Neomphalioidea, with three families, of which the family Peltospiridae includes twelve genera and twenty-one living species. All species of Peltospiridae live in communities near hydrothermal vents, which emit hot, acidic water rich in minerals and toxins such as hydrogen sulfide. These communities rely on chemosynthetic energy production by bacteria that derive energy from dissolved chemicals such as hydrogen sulfide to fix carbon into organic matter, unlike most communities that are based on energy from sunlight (photosynthesis).

Some genera of Peltospiridae, such as *Chrysomallon* and *Gigantopelta*, have comparatively large shells (> 2 in / > 50 mm) consisting of three to four broad, globose whorls with a low spire and a large oval aperture. The shells have a thick periostracum and a multispiral operculum.

Others, such as *Pachydermia*, have minute shells ($1/16$–$3/16$ in / 2–5 mm) that have a higher spire, with the later whorls with open coiling.

DISTRIBUTION
Global, found in deep-sea hydrothermal vents on the ocean floor.

DIVERSITY
Family includes 22 living species assigned to 13 genera within 1 subfamily.

HABITAT
Occur in high densities along hydrothermal vents at abyssal depths (6,560 ft / 2,000 m) and are adapted to sulfide-rich reducing habits.

SIZE
Species range in size from $1/16$–$1\frac{3}{4}$ in (2–45 mm).

DIET
Peltospirids such as *Chrysomallon* obtain all their nutrition from endosymbiotic chemoautotrophic bacteria. They do not feed. Others may feed on layers of detritus and bacteria on rocks.

Species in other genera (such as *Ctenopelta*, *Rhynchopelta*, and *Nodopelta*) are shaped more like limpets or abalones with a rapidly enlarging aperture and very low spire. Some lack an operculum.

Animals have large cephalic tentacles without eyes, a long, tapered snout, a rhipidoglossate radula, a single left bipectinate gill and a single left kidney, a heart with a single auricle, and a rectum that passes ventral to the heart. Some taxa maintain chemosynthetic bacteria in their gills; others (for example, *Chrysomallon*) maintain them in an enlarged, modified, and highly vascularized portion of the esophageal gland. The sides of the foot of *Chrysomallon squamiferum* (Sea Pangolin) are clad in scales. The shell as well as the scales contain iron sulfide. This is the only animal known to incorporate iron sulfide into its shell. The family Peltospiridae has been reported from a fossilized chemosynthetic community dating from the Middle Eocene.

OPPOSITE | *Pachydermia sculpta*, a species that occurs near hydrothermal vents at depths of 5,740–6,560 ft (1,750–2,000 m) in basins around Fiji.

ABOVE | *Chrysomallon squamiferum* is endemic to hydrothermal vents in the Indian Ocean. Found at depths of 7,875–9,500 ft (2,400–2,900 m).

REPRODUCTION
Sexes are separate for most studied species, except for *Chrysomallon squamiferum*, which is a simultaneous hermaphrodite. Larvae develop from large, yolky eggs dispersed over large distances by bottom currents.

PLEUROTOMARIIDA—PLEUROTOMARIOIDEA—PLEUROTOMARIIDAE
SLIT SHELLS

Members of the family Pleurotomariidae are easily recognized by their large, conical, spirally coiled shells with a nacreous inner layer and a characteristic slit along the outer lip that produces the selenizone, a distinctive band that extends from the rear of the slit along the earlier whorls of the shell. The mantle edge along the slit is lined with papillae that may interdigitate except near the posterior margin of the slit, providing an exhalant opening to a functionally tubular mantle cavity. Within the mantle cavity, the slit is flanked by paired mantle cavity organs (e.g. gills, osphradia, hypobranchia glands, and kidneys) that are larger on the left side than on the right.

Pleurotomariids were common in shallow-water faunas during the Mesozoic but thought to be extinct until living specimens were discovered in the deep sea in the mid-nineteenth century.

Based on limited observations, the five genera of pleurotomariids appear to segregate by depth, while its forty-two species segregate geographically, with minimal overlap in their ranges. Two of the five genera (*Entemnotrochus* and *Bayerotrochus*) are present in both the Atlantic and Indo-Pacific. The genus *Bouchetitrochus* ranges from Japan to the vicinity of New Caledonia, while *Mikadotrochus* inhabits waters from Japan through the Philippine Islands.

Unlike most other Vetigastropoda, which are herbivores, pleurotomariids are carnivorous, feeding predominantly on sponges and soft corals. Their alimentary systems

BELOW | A dorsal view of *Entemnotrochus rumphii*, showing the extent of the slit.

DISTRIBUTION
Along the temperate and tropical western margins of the Atlantic and Indo-Pacific Oceans, including Japan and the Philippines.

DIVERSITY
Family includes 42 living species assigned to 5 genera within 1 subfamily.

HABITAT
Hard substrates at bathyal depths ranging from 330–3,300 ft (100–1,000 m).

SIZE
Species range in size from 1–8 in (30–200 mm).

DIET
Primarily sponges and soft corals.

REPRODUCTION
Eggs are shed and fertilized externally. The free-swimming larval stage is nourished by a large yolk.

ABOVE LEFT | *Bayerotrochus midas* on rock wall at 1,400 ft (427 m), Portales Terrace, off the Florida Keys.

are highly adapted for this diet, with specialized radular teeth and a chitin-lined buccal cavity to protect it from sponge spicules.

Predators include crustaceans and fish. The hypobranchial glands of pleurotomariids rapidly secrete large amounts of viscous white fluid—thought to repel predators—when the animals are disturbed.

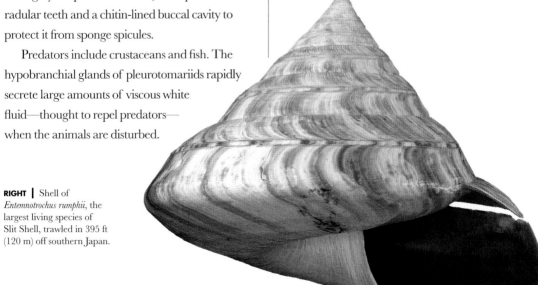

RIGHT | Shell of *Entemnotrochus rumphii*, the largest living species of Slit Shell, trawled in 395 ft (120 m) off southern Japan.

LEPETELLIDA—HALIOTOIDEA—HALIOTIDAE
ABALONE

Haliotidae are easily recognized by their distinctive ear-shaped shell, with a very large, oval aperture that is strongly tilted relative to the coiling axis of the shell. This results in a low spiral structure consisting of only two or three whorls.

The flattened shell also has a series of holes (tremata) that are diagnostic of the family. New tremata are formed at the growing edge of the shell. The posterior tremata are sealed so that only the youngest five to ten remain open at any time. Water enters the mantle cavity through the anterior holes and exits through the posterior holes, carrying wastes as well as gametes during spawning. The outer shell surface usually has spiral ribs and commarginal cords and tremata may have short, tubular outgrowths that form a spiral row of knobs when they are sealed. The inner shell surface is composed of nacre (mother

RIGHT | Anterior view of the animal of *Haliotis asinina*. Tremata along the surface of the shell are visible, as are the large papillae along the mantle edge.

DISTRIBUTION
Primarily temperate latitudes but can occur globally in the tropics.

DIVERSITY
Family includes 57 living species assigned to 1 genus.

HABITAT
Rocky bottoms, mostly at shallow subtidal depths (65 ft / 20 m), with some species living in intertidal areas and others at bathyal depths of 400–2,400 ft (130–1,200 m).

SIZE
Species range in size from ¾–12 in (20–300 mm) in length.

DIET
Macroalgae, preferring red or brown algae. Most are nocturnal feeders.

REPRODUCTION
Sexes are separate. Fertilization is external. Veliger larvae are lecithotrophic, nourished by large yolks, and remain in the plankton for about 5 days before settling to the bottom.

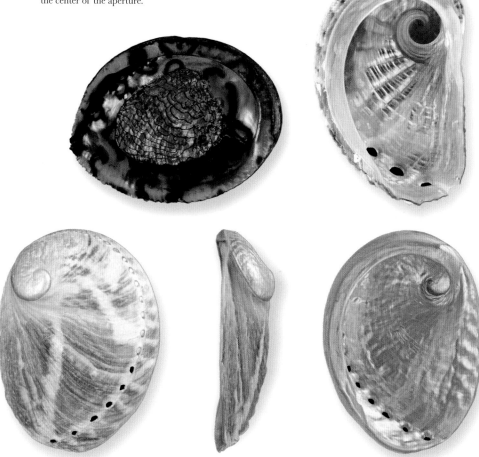

BELOW | The inner surface of the shell of *Haliotis fulgens* with the conspicuous large muscle scar at the center of the aperture.

RIGHT | A ventral view of the shell of *Haliotis elegans*, showing the earlier whorls.

ABOVE | Dorsal, lateral, and ventral views of the shell of *Haliotis laevigata*, a common shallow-water species occurring off the coast of southern Australia.

of pearl), which may be white or iridescent with multiple colors.

The large muscle by which the animal is attached is near the center of the shell and may produce a conspicuous muscle scar. An operculum is absent.

The animal retains paired mantle cavity organs on each side of the spiral row of tremata. The mantle edge is lined with multiple, often elaborate papillae extending beyond the shell's edges.

The foot is large, oval, and muscular, with an anterior notch through which the snout is extended when feeding. The rhipidoglossan radula is used to scrape and shred algae.

Abalones are harvested commercially throughout the world, and many species are considered threatened. Several have protected status in at least part of their geographic ranges. Earliest records of the family Haliotidae are from the latest Upper Cretaceous (Maaestrichtian 66–72 mya) deposits.

LEPETELLIDA—FISSURELLOIDEA—FISSURELLIDAE
KEYHOLE LIMPETS

Fissurellidae are called Keyhole Limpets because the foramen, which sits at the apex of many species' shells, resembles a keyhole. Some genera have a slit or notch at the anterior shell margin; others entirely lack an opening. Despite their limpet-like shells, fissurellids are more closely related to gastropods with coiled shells such as Trochidae than to true limpets.

Shells range from nearly flat to steeply conical, with the apex curving posteriorly. The surface may be smooth or sculptured to varying degrees with commarginal bands and radial ribs that tend to form a cancellate pattern with nodules or scales where they intersect. Shell color ranges from white to a variety of colors, mostly shades of brown or red, often in radial patterns. The interior shell surface is porcelaneous with a horseshoe-shaped muscle scar open at the anterior end. The foramen edges are resorbed and redeposited as the shell grows. The protoconch consists of a single coiled whorl. An operculum is present in larvae but lost at metamorphosis.

Animals have a large foot, a blunt snout, stout cephalic tentacles with eyes on short stalks at their bases, as well as complex and distinctive radula with numerous teeth per transverse row. The fissurellid mantle cavity is similar to that of primitive vetigastropods in being bilaterally symmetrical

DISTRIBUTION
Global, primarily in warm to temperate latitudes and shallow waters, but also present in polar regions and at abyssal depths.

DIVERSITY
Family includes approximately 500 living species assigned to 48 genera within 5 subfamilies.

HABITAT
Hard substrates, including rocks, rubble, and kelp stems, mostly at depths sufficiently shallow to support algal growth, though some occur at depths of over 6,650 ft (2,000 m).

SIZE
Species range in size from ¼–5 in (6–132 mm).

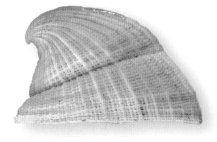

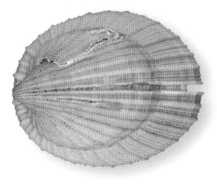

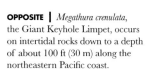

and containing paired gills, osphradia, hypobranchial glands, and kidneys. In species with an apical foramen, the mantle may form a thickened exhalant siphon. In some species, the mantle extends nearly completely over the shell. Life spans between two and fifteen years have been reported for several species.

Early Fissurellidae fossil records date from the Middle Triassic (Ladinian 232–147 mya). The blood of *Megathura crenulata* contains hemocyanin, a copper-based respiratory pigment used in the production of vaccines and for treating several cancers in humans.

LEFT | *Emarginula tuberculosa*, a fissurellid with an anterior slit rather than a foramen at the apex of the shell.

BELOW | *Diodora italica*, a common shallow-water fissurelid from the northern Mediterranean.

OPPOSITE | *Megathura crenulata*, the Giant Keyhole Limpet, occurs on intertidal rocks down to a depth of about 100 ft (30 m) along the northeastern Pacific coast.

DIET
Algae, kelp, and seagrasses, forams, hydrozoans, bryozoans, nematodes, tunicates, and detritus. Some species feed on sponges.

REPRODUCTION
Sexes are separate. Gametes are released through the foramen and fertilization is external. Larvae are pelagic and lecithotrophic, relying on yolk for nutrition. Some species have short larval spans of only a few days.

TROCHIDA—TROCHOIDEA—TROCHIDAE
TOP SHELLS

Top Shells are among the more common snails inhabiting rocky intertidal shores along many coastlines throughout the world, although some occur on soft sediments and at all ocean depths.

Many have tall, conical shells composed of multiple, tightly coiled whorls, with a base that is flat or rounded. Others have low, broad shells, or shells with few, rapidly expanding whorls. The aperture is ovate, generally elongated toward the outer shell margin, and without a siphonal canal. The outer lip is oriented at an angle to the coiling axis of the shell and may be thin or thick, smooth, or with lirae along its inner edge. The columella may be rounded and smooth or reinforced with multiple spiral lirae. The shell exterior may be smooth, or have spiral cords, and/or axial ribs, some with nodules or tubercles. The periostracum may be thin or thick. The inner surface of the shell is nacreous. The operculum is corneous, round, multispiral, and fits tightly within the aperture.

Unlike the more primitive families within Vetigastropoda, Trochidae no longer have paired organs within the mantle cavity, having lost their right gill, osphradium, and kidney due to narrowing of the aperture resulting from tighter shell coiling. The foot is broad and has two to three pairs of epipodial tentacles and sense organs. The head has a short snout, a pair of tentacles, and eyes at the ends of stalks. Most Top Shells use their radula to scrape algae and detritus from hard substrates, but species of the genus *Umbonium* are filter feeders.

Their fossil record extends to the Middle Triassic (Ladinian 242–237 mya).

LEFT | *Umbonium vestiarium*, an abundant species found on shallow-water mud bottoms throughout the Indo-West Pacific.

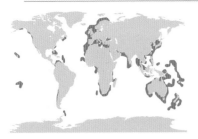

DISTRIBUTION
Global, ranging from the tropics to both polar regions.

DIVERSITY
Family includes approximately 500 living species assigned to 76 genera within 11 subfamilies.

HABITAT
The majority of genera and species inhabit hard rocky shores and reefs at intertidal and subtidal depths. Some genera live at bathyal and abyssal depths.

SIZE
Species range in size from ¼–6 in (6–150 mm).

RIGHT | Dorsal surface of the shell of *Clanculus puniceus*, a common intertidal species living under rocks off the eastern coast of Africa.

BELOW RIGHT | *Gibbula fanulum* is a small species limited to Portugal's southern coast and the western Mediterranean Sea.

ABOVE | *Steromphala cineraria* on a kelp leaf off the coast of Norway. Cephalic tentacles as well as epipodial tentacles are visible around the edges of the shell.

BELOW LEFT | An apical view of the iridescent shell of *Gaza fischeri*, a deep-water species from the tropical western Atlantic. Some species of *Gaza* have been observed to swim short distances by undulating their broad foot to escape predators.

BELOW | Apertural, ventral, and lateral views of the shell of *Tectus dentatus*, a subtidal species that occurs in the Red Sea and the northwestern Indian Ocean.

DIET
Most species are herbivorous grazers, feeding by scraping algae and detritus from the substrate. Several species are filter feeders.

REPRODUCTION
Sexes are separate. Fertilization is external. Some species lay eggs in gelatinous masses. Eggs hatch as planktonic larvae.

TROCHIDA—TROCHOIDEA—TURBINIDAE
TURBAN SHELLS

Turban Shell species are among the largest and most abundant herbivorous snails inhabiting the lower intertidal zone of many tropical shorelines. This family is related to Trochidae, from which it is most easily distinguished by the presence of a calcified operculum (corneous in Trochidae) as well as by significant differences in their radular teeth, which require dissection and microscopic examination to discern.

Most turbinid shells are large, thick, conical, or turbinate with rounded whorls, although some genera have low, broad shells. The aperture is round to ovate, nacreous, and lacks a siphonal canal. The columella is rounded and smooth, without lirae. The shell exterior may be smooth and glossy, or have spiral cords with nodules, scales, or open spines along the shoulder. The operculum fits the shell aperture tightly and is thick and heavily calcified externally. Its inner surface may show spiral coiling.

The animal has a large, ovate, muscular foot with epipodial tentacles along its sides, a head with a short snout that is split mid-ventrally, and eyes on stalks at the base of long tentacles. Like trochids, they no longer have paired, symmetrical mantle cavity organs such as gills or kidneys, having lost these organs on the right side.

Earliest fossil records of Turbinidae are from deposits dated to the late Permian (Wordian 266.9–264.2 mya), and are older than the oldest Trochidae from the Middle Triassic (Ladinian 242–237 mya), suggesting that these two lineages diverged by the early Mesozoic.

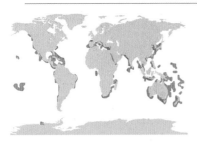

LEFT | *Turbo petholatus* occurs throughout the Indo-West Pacific Ocean. It has a thickly calcified circlular operculum that is green at its center and commonly called the Cat's Eye.

DISTRIBUTION
Global, ranging from the tropics to polar regions in all oceans, with the majority of species inhabiting shallow waters in tropical and subtropical latitudes.

DIVERSITY
Family includes approximately 160 living species assigned to 17 genera within 2 subfamilies.

HABITAT
Hard substrates as well as sandy bottoms, especially around coral reefs, at depths ranging from intertidal to bathyal (0–1,640 ft / 0–500 m).

SIZE
Species range in size from ½–9 in (12–120 mm).

RIGHT | *Guildfordia yoka* has long radial spines that triple its effective diameter, making it difficult for predators to overturn. It inhabits deep waters off Japan.

BELOW | Apical and ventral views of the green color form of *Turbo petholatus*.

BELOW | *Bolma girgyllus* inhabits offshore depths of 165–330 ft (50–200 m) along the tropical western Pacific Ocean. Its long, delicate spines may be covered by encrusting organisms.

DIET
Herbivores, feeding on algae, seagrasses, and detritus.

REPRODUCTION
Sexes are separate. Fertilization is external. Eggs are released in gelatinous masses and hatch as planktonic larvae.

LEPETELLIDA—LEPETELLOIDEA—PSEUDOCOCCULINIDAE
WOOD LIMPETS

BELOW | *Notocrater youngi*, living on a sunken wooden branch collected in 980 ft (300 m) of water off Dominica.

The Pseudococculinidae is one of several families of small snails with limpet-like shells that live in the deep sea, some along hydrothermal vents, each adapted to life on one or more biogenic substrates such as sunken wood, algal holdfasts, whale bones, shark egg cases, carapaces of deep-sea crabs, or squid beaks that sink to the seafloor. As the name Pseudococculinidae implies, the shells of the limpets in this family very closely resemble those of the family Cocculinidae, although these families are unrelated, and each has independently evolved from a coiled ancestor to a limpet-like shell.

Pseudococculinidae is the most species-rich of the deep-sea limpet families. Most species live on sunken wood, although some live on crab carapaces; others along hydrothermal vents.

Pseudococculinid limpets have small, conical, oval shells with a subcentral or posteriorly directed apex. The protoconch has a distinctive, long, and narrow fold, but varies in surface sculpture. The shell sculpture consists of commarginal growth lines or bands, some with pronounced nodules in radial patterns.

DISTRIBUTION
Global, at bathyal to abyssal depths.

DIVERSITY
Family includes 40 living species assigned to 11 genera.

HABITAT
Live on a variety of biogenic substrates, primarily wood and algal holdfasts that sink to the ocean floor, but also on carapaces of deep-sea crabs.

SIZE
Species range in size from 1/16–3/4 in (1½–17 mm).

DIET
Most species feed on bacterial films growing on and in degraded sunken wood or on other biogenic substrates.

Animals have a muscular foot, broad oral lappets, laterally situated cephalic tentacles (the larger right tentacle is modified to form a copulatory organ), and a pair of posteriorly directed epipodial tentacles. The genera of Pseudococculinidae as well as the various families of deep-sea limpets are most easily identified and distinguished on the basis of significant differences in anatomy and the morphology of their radular teeth.

ABOVE | *Tentaoculus granulatus*, a deep-water species living along methane seeps off the mouth of the Congo River, western Africa.

BELOW | *Tentaoculus haptricola*, from sunken algal holdfasts collected at bathyal depths off New Zealand.

REPRODUCTION
Simultaneous hermaphrodites. Fertilization is internal. Eggs are produced in large numbers and larvae can delay metamorphosis until they encounter an appropriate substrate.

CYCLONERITIDA—NERITOIDEA—NERITIDAE
NERITES

Members of the family Neritidae are among the most abundant gastropods that inhabit intertidal hard substrates such as rocks, seawalls, pilings, and mangroves along tropical to temperate coastlines around the world. In many areas, several species can co-occur in sympatry. Many of the genera have adapted to brackish and freshwater habitats.

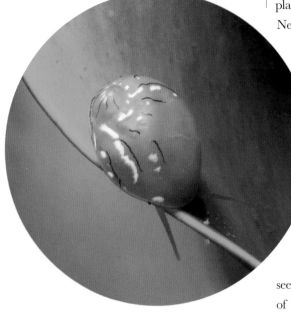

BELOW | *Smaragdia viridis* crawling on seaweed off Tenerife, Canary Islands.

Shells are thick, rounded, and nearly spherical in shape. Surface sculpture varies from smooth, with simple growth lines, to prominent spiral ribs. Colors may be fairly uniform or with elaborate patterns. The columella is thickened to form a shelf-like septum across the aperture. Both the septum and the inner surface of the outer lip may have folds or teeth. Animals can withdraw completely into the shell and seal the aperture with a thick, calcified, D-shaped operculum with an internal peg (apophysis) that helps it to completely seal the aperture and lock into place. This offers protection from desiccation. Neritids can escape predators by retracting into their rounded shells and falling or rolling down steep, rocky shores.

Neritid animals have a body that is more limpet-like than coiled, with a broad foot; their head has a pair of tentacles with eyes at the base. Their radula is rhipidoglossate. They lack salivary glands and a true jaw and have only a left gill and left kidney. Some freshwater and marine species may serve as intermediate hosts for trematode parasites. *Bathynerita naticoides* is a recently discovered species endemic to deep-water cold seeps in the Gulf of Mexico and Caribbean at depths of 1,310–6,890 ft (400–2,100 m). It lives on the shells

DISTRIBUTION
Tropical, subtropical, and temperate latitudes around the world. Some genera inhabit estuaries; others occur in freshwater habitats.

DIVERSITY
Family includes approximately 175 living species assigned to 16 genera within 2 subfamilies.

HABITAT
Common on hard substrates such as rocky shores and mangrove roots, mostly at intertidal and shallow subtidal depths.

SIZE
Species range in size from 5/16–1½ in (8–38 mm).

DIET
Omnivores, detritivores, or herbivores, grazing on algae, diatoms, and detritus.

LEFT | *Nerita textilis* inhabits rocky shores at or above the high-tide line. It relies on water splashed by waves and is capable of surviving out of the water for long periods of time.

BELOW | *Vittina waigiensis* inhabits coastal mangrove forests and can survive in saltwater and brackish water habitats.

of vent mussels, feeding on detritus on their surfaces as well as on their decomposing periostracum and byssal fibers.

Earliest fossils date from the middle Devonian (398–385 mya).

Some freshwater species are carnivorous, feeding on insect larvae.

REPRODUCTION
Sexes are separate. Fertilization is internal. Males produce spermatophores that facilitate sperm transfer. Females deposit pale gelatinous eggs either singly or in groups in egg capsules on rocks. Eggs hatch as planktonic larvae. Some freshwater species have direct development.

CAENOGASTROPODA—CERITHIOIDEA—CERITHIIDAE
CERITHS

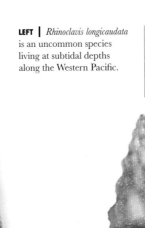

LEFT | *Rhinoclavis longicaudata* is an uncommon species living at subtidal depths along the Western Pacific.

Members of the family Cerithiidae, commonly known as Ceriths, are often encountered on sheltered intertidal sand and mud flats along tropical and subtropical shores and in estuaries, where they occur in large numbers either on or partially buried in the substrate. Several genera of smaller Ceriths live in deeper waters offshore, at depths of up to 3,280 ft (1,000 m).

The shell is elongated and high-spired, consisting of multiple, closely spaced whorls that may be straight-sided, slightly convex, or shouldered, some with irregularly spaced varices, indicating periodic interruptions in shell growth. The aperture is broadly oval, tilted relative to the coiling axis of the shell, with a siphonal canal at the anterior margin. The outer lip may be thin, undulating, or thickened and slightly flaring. The columella

LEFT | *Cerithium citrinum* occurs along the shores of Northeastern Australia and Melanesia.

DISTRIBUTION
Tropical, subtropical, and temperate latitudes worldwide.

DIVERSITY
Family includes 200 living species, assigned to 26 genera and 2 subfamilies.

HABITAT
Primarily shallow water, marine and estuarine sand, mud, or rubble bottoms. Several genera live offshore at depths of up to 3,280 ft (1,000 m).

SIZE
Species range in size from ⅛–6 in (3–150 mm).

DIET
Grazing on the seabed, feeding on algae and detritus.

REPRODUCTION
Sexes are separate. Eggs are deposited in gelatinous strings attached to substrate, with larvae hatching as plankton.

ABOVE | *Cerithium caeruleum* feeding on marine algae.

is smooth and may be thickened at its posterior margin, defining an exhalant canal.

The shell surface may have spiral cords and axial ribs, with or without nodules. Shell color ranges from white to dark brown and may be marked with regular or irregular bands of darker color. The operculum is thin, corneous, and paucispiral.

Animals have a broad, short foot, a small anterior siphon, a mantle edge with papillae and long tentacles with eyes at their bases.

The snout is long and broad, adapted for foraging for algae and detritus.

The oldest Cerithiidae are recognized in fossil deposits from the Early Cretaceous (Albian 113.0–100.5 mya).

CAENOGASTROPODA—CERITHIOIDEA—TURRITELLIDAE
TURRET SNAILS

Turret Snails comprise a lineage that has become highly specialized and adapted for filter feeding on subtidal sand bottoms, primarily in tropical and subtropical regions, preferring cooler rather than warmer waters.

Most turritellid shells are tall, narrow, and high-spired, with numerous closely spaced whorls and without an umbilicus. The aperture is small and may be rounded or rectangular, with a columella that is smooth and rounded, and an outer lip that is thin and may be rounded or shouldered and angled relative to the coiling axis of the shell. There is no siphonal canal. The shell surface may be smooth, or covered with spiral threads or cords that may have nodules, depending on species. The operculum is thin, multispiral, with a central nucleus, and may have bristles along its border. Animals have a small foot, a large head with tentacles and eyes, and a mantle edge that is frilled. The mantle cavity is deep and contains a long gill that filters food particles from the water. Some species supplement their diet by deposit feeding.

BELOW | *Turritella bacillum* partially buried in sand, in shallow water off the coast of India.

DISTRIBUTION
Global, primarily in tropical and temperate latitudes, but extending to polar regions.

DIVERSITY
Family includes approximately 150 living species assigned to 18 genera within 5 subfamilies.

HABITAT
Sand rubble and mud bottoms at depths ranging from intertidal to nearly 4,920 ft (1,500 m). Many are mobile, but species in the subfamily Vermiculariinae are sessile and can form reefs of entangled shells or become embedded in sponges or ascidians.

SIZE
Species range in size from 1–6½ in (25–165 mm).

LEFT | *Vermicularia spirata* is a common tropical western Atlantic species that may become partially embedded in sponges or colonial ascidians.

Species in the subfamily Vermicullariinae begin life like other turritellids, producing tightly coiled shells for the first few whorls, then becoming unwound and growing in irregular drawn-out worm-like patterns. They are protandric hermaphrodites and begin life as males that are mobile and capable of finding females while their shells are tightly coiled. As they grow, they become females, their irregularly formed shells becoming intertwined with those of other females, forming colonies of varying sizes that continue to attract mobile males.

Fossils with similar shells have been reported from as early as the Devonian, but Turritellidae can be traced to the Early Cretaceous (Valanginian 139.8–132.9 mya) with reasonable confidence.

LEFT | *Turritella terebra*, the largest species in the genus, is abundant on subtidal sandy bottoms throughout the Indo-West Pacific.

DIET
Adapted for subtidal filter feeding on soft bottoms.

REPRODUCTION
Sexes are separate in most species. Some are protandric hermaphrodites. Males transfer spermatophores. Eggs are deposited in capsules attached to hard substrate and hatch as veliger larvae or are brooded within the mantle cavity of the female and hatch as crawling young.

CAENOGASTROPODA—CERITHIOIDEA—SILIQUARIIDAE
SPONGE WORM SHELLS

Sponge Worm Shells are adapted to a sessile life embedded in sponges and coralline algae. They generally occur as aggregations of several individuals in a sponge. Early whorls of the shell are tightly coiled, but after the first few, the whorls become unwound, although maintaining an irregular dextral coiling direction, with only the aperture protruding slightly above the outer surface of the sponge. Sculpture is generally limited to spiral cords that may have scales or spines. Some genera have a longitudinal slit along the dorsal surface of the shell that may be periodically closed to form a series of holes. Some shells may have septa sealing off earlier parts of the shell. The aperture is round. The operculum is flat or conical, with long bristles that may be branched.

They have a short, blunt snout, a short pair of cephalic tentacles with small eyes at their bases and a radula with seven teeth per transverse row. The mantle edge has short papillae. Siliquariidae are suspension feeders, with a deep mantle cavity containing a long gill with elongated filaments that

DISTRIBUTION
Global, in tropical and temperate latitudes, at subtidal to bathyal depths.

DIVERSITY
Family includes 38 living species assigned to 5 genera in 1 subfamily.

HABITAT
Live in groups within sponge hosts, with the apertures extending beyond the sponge's surface. Occur in 35 different species of sponges, at depths ranging 33–1,440 ft (10–440 m).

SIZE
Species range in size from 2–18 in (50–450 mm).

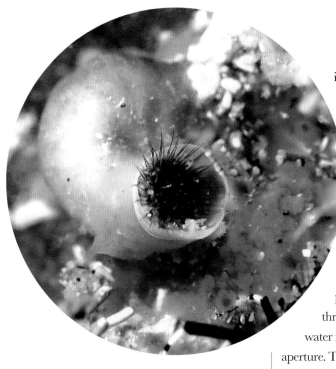

incorporate food particles filtered from the water into a mucus rope that is transferred to a food groove along the side of the foot and then to the mouth by a special finger-like pre-oral projection.

In species with a slit, water enters at the aperture on the sponge's surface and exits through the slit or holes after it is filtered. The operculum is periodically used as a plunger to flush the mantle cavity through the slit. In species without a slit, water is discharged on the opposite side of the aperture. The earliest report of Siliquariidae is from the Middle Triassic (237 mya).

OPPOSITE AND BELOW | Five views of a shell of *Tenagodus anguinus* from the Indian Ocean, with dorsal slit closed to form a series of holes on the upper whorls.

ABOVE | *Stephopoma roseum* growing on underside of encrusted rock in the intertidal zone off northern New Zealand.

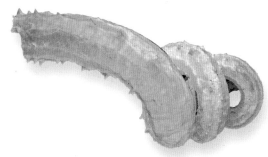

DIET
Filter feeders.

REPRODUCTION
Sexes are separate. May be protandric hermaphrodites. Males are usually aphallic and shed spermatophores that enter aperture of neighboring females. Fertilization is internal. Brood their young either freely within the mantle cavity or in specialized brood pouches on the right side of the head-foot.

LITTORINIMORPHA—LITTORINOIDEA—LITTORINIDAE
PERIWINKLES

Periwinkles, as littorinids are commonly known, are conspicuous and abundant inhabitants of intertidal rocky shores. Some genera live on mangrove roots and seagrasses throughout the world; others live on subtidal seaweed such as kelp. Multiple species co-occur in some areas, and show both vertical and horizontal patterns of zonation, but may vary in the degree to which their ranges overlap. Many species living near the high-tide line are exposed to air for more time than they are submerged. Others occur above the high-tide line, relying on spray from breaking waves, and tend to be most active at night and when it rains.

Littorinid shells span a variety of shapes and sizes. Most are thick and fairly heavy, with a conical spire of few whorls and a broadly rounded anterior margin formed by a large, nearly circular aperture with a smooth, thickened columella and lacking a

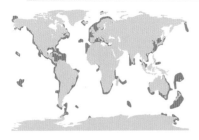

DISTRIBUTION
Global, being most diverse in tropical and temperate regions but also occurring in polar regions.

DIVERSITY
Family includes approximately 200 living species assigned to 18 genera and 3 subfamilies.

HABITAT
Occupy a variety of habitats including rocky shores from above the high-tide line to subtidal depths of 150 ft (50 m), where some live on seaweed. Others live on mangroves and seagrasses.

SIZE
Species range in size from ⅛–2½ in (3–65 mm).

siphonal canal. The outer surface may be smooth, or have spiral cords, that in some genera form axially aligned nodules. Shell colors vary from whitish to brown, many with darker spiral or axial bands. The inner surface is smooth and porcelaneous. The operculum is thin, corneous, and tightly fits the aperture.

The animal has a large, muscular foot with a median furrow. The head has a short broad snout and two tapering tentacles with eyes along their outer base. The mantle cavity contains a single gill and is vascularized to function as a lung.

A single species, *Cremnoconchus conicus*, inhabits fresh water in western India. All other littorinids are marine.

The oldest records of littorinids are from the Late Cretaceous (Coniacian 89.8–86.3 mya).

OPPOSITE | *Littorina littorea* on a frond of seaweed in the White Sea, Karelia, Russia.

ABOVE | *Littorina obtusata* on kelp at low tide, Isle of Islay, Scotland.

BELOW | Dorsal and apical views of the shell of *Littoraria angulifera*, a species that lives on mangrove roots above the waterline along the tropical western Atlantic.

BELOW | Dorsal and apical views of the shell of *Tectarius coronatus*, a species common on intertidal rocks in the islands of the western Pacific Ocean.

BELOW | Dorsal and lateral views of the shell of *Littorina compressa*, a species that inhabits rocky shorelines along the northeastern Atlantic.

DIET
Herbivores, feeding on diatoms and algae scraped from hard substrates.

REPRODUCTION
Sexes are separate. Fertilization is internal. Eggs may be laid individually, each in a floating capsule, shed directly into the sea, or deposited in gelatinous masses that hatch as free-swimming planktonic larvae. Some species brood eggs in the mantle cavity until they hatch.

LITTORINIMORPHA—TRUNCATELLOIDEA—CAECIDAE
CAECUM SHELLS

The Caecidae are a family of minute snails, usually only a few millimeters in length, that occur mostly in shallow tropical and temperate seas around the world. They occupy a variety of habitats occurring epifaunally among algae, seagrasses, and corals, as well as interstitially in pore spaces between sediments.

BELOW | *Caecum californium* crawling on seaweed in a tide pool off La Jolla, California.

Two subfamilies are generally recognized. Species of genera included in Caecidae pass through three morphologically distinct growth stages. Larval shells consist of two coiled whorls. After the larvae settle to the bottom, they begin to form the adult shell, which becomes uncoiled as it grows larger. Portions of the shell, at first the protoconch and then one or more posterior sections, are shed as the shell continues to grow and become progressively less coiled. Each time the opening is sealed with a septum, which may be smooth or have a finger-like projection called a mucro. The adult shell is short and tubular and may be smooth or sculptured with commarginal bands.

Members of the subfamily Ctiloceratinae have a protoconch that is initially coiled but becomes a straight or slightly curved tube. After metamorphosis, the protoconch is retained and the adult shell (teleoconch) is planispirally coiled, with a wide and deep umbilicus. Other genera that retain their protoconch but have an uncoiled adult shell are generally included in Ctiloceratinae. Members of all genera have an operculum that is circular, thick, corneous, and multispiral, with a central nucleus.

DISTRIBUTION
Global, at tropical and temperate latitudes.

DIVERSITY
Family includes approximately 260 living species assigned to 10 genera within 2 subfamilies.

HABITAT
Most are epibenthic in algae or corals, or interstitial in shallow-water marine sediments, but some species occur at depths to 3,280 ft (1,000 m).

SIZE
Species range in size from 3/16–1/16 in (2–5 mm).

DIET
Microphagous grazers.

REPRODUCTION
Sexes are separate. Eggcapsules are spherical and covered with detritus. Larvae hatch as pelagic veligers.

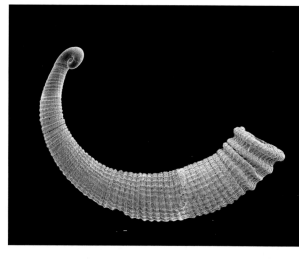

LEFT | *Parastrophia japonica*, a species from southern Japan that retains its coiled protoconch at the tip of a slightly curved shell.

BELOW LEFT | Dorsal and right lateral views of *Caecum gracile*. Earlier shell portions have been shed and the posterior opening sealed with a septum.

ABOVE | *Caecum floridanum*, reassembled to show sections that are shed as the animal grows.

Animals have a short, narrow foot, a short snout, and long, cylindric cephalic tentacles with eyes at their bases.

Taxonomy of members of Caecidae is currently based primarily on shell morphology and continues to be revised as more anatomical and molecular data is accumulated. Most of the ten currently recognized genera have few species and are limited in distribution to the Indian and Pacific Oceans. The genus *Caecum*, which includes over 200 species, has a circumglobal distribution throughout tropical and temperate seas.

Earliest fossil records of Caecidae are from the late Paleocene (Thanetian 59.2–56 mya).

LITTORINIMORPHA—STROMBOIDEA—STROMBIDAE
CONCHES

The term "conch" is usually applied to members of the family Strombidae but has also been used to refer to other large, edible gastropods in some regions.

Shells of Strombidae vary greatly depending on genus, ranging from moderately sized, fusiform, and high-spired, to very large, massive shells with thick, flared outer lips. The genus *Lambis* has shells characterized by an expanded outer lip with multiple broad, outwardly directed spines, while the genus *Tibia* has shells that are extremely high-spired and narrow, with a small aperture, short digitations, and a thin, narrow, axial siphonal canal that is nearly half the length of the shell. All have a stromboid notch, a

DISTRIBUTION
Global, in mostly shallow tropical to subtropical marine habitats with greatest diversity in the Indo-Pacific. Many genera are confined to the tropical Indo-Pacific.

DIVERSITY
Family includes approximately 120 living species assigned to 30 genera within 1 subfamily.

HABITAT
Most species inhabit nearshore sandy bottoms, seagrass meadows, and coral reefs in shallow waters ranging from intertidal to depths of less than 200 ft (61 m). Some genera, such as *Tibia*, live on muddy bottoms in deeper waters (20–500 ft / 6–152 m).

SIZE
Species range in size from ¾–16 in (19–400 mm).

FAR LEFT | *Latissistrombus sinuatus*, a shallow-water species common throughout the southwestern Pacific.

LEFT | *Tibia fusus* occurs in deeper waters in the southwestern Pacific. Some researchers have segregated this genus to a separate family, Rostellariidae.

diagnostic indentation near the anterior end of the outer lip that accommodates the right peduncle, one of two broad stalks that bear a prominent eye, another distinguishing family feature.

Animals have a long, muscular snout, with a short, broad radular ribbon. The mantle contains a well-developed gill, and a long, narrow osphradium. The foot is long, narrow, and muscular with a claw-like operculum that may have a serrated edge. Strombids move by anchoring the foot in the substrate and moving in a series of vaulting motions, lurching forward in small increments. They undergo determinate growth; the shell grows rapidly until maturity, at which time the outer lip becomes thickened, flared, or otherwise modified. They may then become thicker and heavier with age.

The oldest records of Strombidae are found in Upper Cretaceous deposits (Coniacian 86.3–89.8 mya) and the family diversified rapidly in the early Cenozoic, achieving maximum diversity during the Eocene. Several species are commercially harvested for food, and for their ornamental shells. Many strombids are gregarious and can occur in small to large colonies.

OPPOSITE | The Queen Conch, *Aliger gigas*, crawling along the seabed in the Bahamas. Eyes are prominent at the tips of the peduncles, or eye stalks.

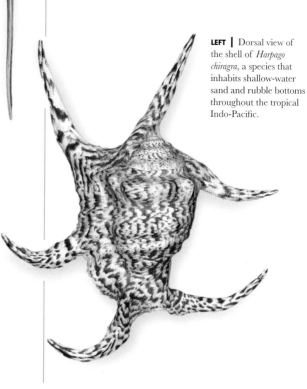

LEFT | Dorsal view of the shell of *Harpago chiragra*, a species that inhabits shallow-water sand and rubble bottoms throughout the tropical Indo-Pacific.

DIET
Herbivores, grazing on algae and seagrass, occasionally consuming detritus.

REPRODUCTION
Sexes are separate; males are smaller than females. Fertilization is internal. Animals may form large mating aggregations. Females deposit eggs in dense gelatinous strands of 100,000 to 500,000 eggs that hatch into planktonic veligers after several days, then settle and metamorphose two to three weeks later.

LITTORINIMORPHA — STROMBOIDEA — APORRHAIDAE
PELICAN'S FOOT SNAILS

The family Aporrhaidae is an ancient lineage with earliest records from the Lower Jurassic (Sinemurian 192.9–199.5 mya) deposits of Europe. It achieved greatest diversity and widest distribution during the Mesozoic, dominating many gastropod faunas and ranging from the tropics to polar regions in all oceans. Many aporrhaids became extinct by the end of the Cretaceous and most others during the late Eocene period. It has been suggested that they were gradually replaced by strombids. The family is represented in the living fauna by two surviving genera. *Aporrhais*, with five species, is limited in distribution to the eastern shores of the Atlantic Ocean from Norway and Iceland to the central coast of west Africa and the Mediterranean. The monotypic genus *Arrhoges* is confined to the northwestern shores of the Atlantic Ocean from Greenland to the coast of the Carolinas.

RIGHT | *Aporrhais pespelecani* crawling along a rubble bottom with head and tentacles extended.

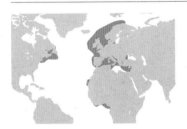

DISTRIBUTION
Shores of the eastern and western Atlantic Ocean and the Mediterranean Sea.

DIVERSITY
Family includes 6 living species assigned to 2 genera within 1 subfamily.

HABITAT
Sand, gravel, and mud bottoms, at subtidal depths 20–1,800 feet (6–550 m), often occurring in large populations. Species may occur in the same geographical region but at different depths and substrates.

SIZE
Species range in size from 1–3 in (25–75 mm).

The shells of aporrhaids have a distinctive shape, with a tall conical spire and a widely flared outer lip with one or more long, finger-like laterally directed extensions. One of the Mediterranean species was said to resemble a pelican's foot, hence the name *Aporrhais pespelecani*. It is among the species discussed by Aristotle.

The animal is yellowish white with red speckles. The head has a long proboscis with a terminal, slit-like mouth and two long tentacles, each with an eye at its base. The foot is narrow and extensile, with a crawling sole and a small, elongated operculum. Aporrhaids move forward by gliding their foot to the anterior margin of the shell then raising the shell and letting it fall forward, advancing in small increments.

ABOVE | Five views of the shell of *Aporrhais serresiana*, a species that occurs at subtidal depths along the eastern Atlantic and in the western Mediterranean.

The animals burrow shallowly into sand or mud bottoms, producing a mound above them and anterior and posterior mucus-lined tubes at the margins of the aperture for water flow. The large gill produces powerful respiratory currents. The animals periodically surface and move short distances to feed every several days before reburrowing.

DIET
Selective deposit feeders, specialized herbivores feeding on diatoms, green algae, and detritus of vegetable origin within and around their burrows.

REPRODUCTION
Sexes are separate. Fertilization is internal. Females deposit soft-walled egg cases in the sand. Larvae hatch as pelagic veligers.

LITTORINIMORPHA—STROMBOIDEA—XENOPHORIDAE
CARRIER SHELLS

Carrier Shells are readily identified by the presence of foreign objects attached to the dorsal surface of their shells. These attached objects vary in size and shape, and, depending on species, may include pebbles, intact or fragmentary shells of other mollusks or brachiopods, coral fragments, and even bottle caps or broken glass. Once the object is in position, the mantle edge produces a secretion to cement the addition, filling gaps with sand particles.

Shells are broadly conical with a low spire and a flat to slightly concave base and rimmed by a peripheral flange of varying width that, together with attached objects, elevates the base above the sediment surface. The dorsal surface may have weak folds or growth striae. One species, *Stellaria solaris*, has characteristic radial spines along its periphery. The aperture is broad and tilted relative to the shell axis with a callused columella. An umbilicus may be present in some species, with a thin periostracum that is often eroded. The shell interior is porcelaneous. The operculum is chitinous with a lateral nucleus and simple growth lamellae.

The animal has a long, narrow muscular foot. The head has a long

LEFT | *Xenophora solarioides* on sand, in shallow water off New Caledonia.

DISTRIBUTION
Global, at tropical and temperate latitudes at depths ranging from subtidal to more than 4,590 ft (1,400 m).

DIVERSITY
Family includes 28 living species assigned to 5 genera within 1 subfamily.

HABITAT
Epifaunal, living on sand and mud bottoms.

SIZE
Species range in size from ¾–6½ in (20–165 mm), although attached objects can more than double the diameter, and epifaunal animals such as sponges and corals can further increase their size.

DIET
Some species feed on detritus, some primarily on foraminifera; others on filamentous algae.

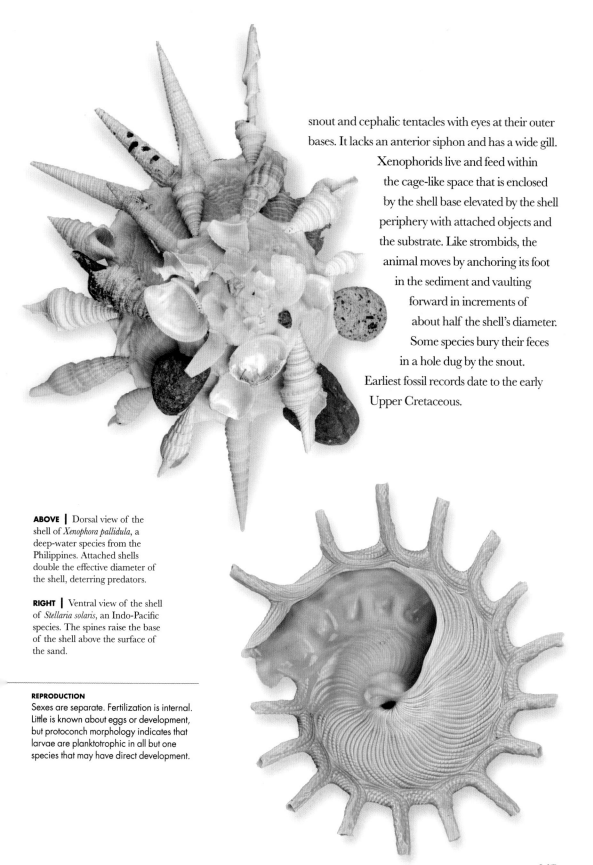

snout and cephalic tentacles with eyes at their outer bases. It lacks an anterior siphon and has a wide gill. Xenophorids live and feed within the cage-like space that is enclosed by the shell base elevated by the shell periphery with attached objects and the substrate. Like strombids, the animal moves by anchoring its foot in the sediment and vaulting forward in increments of about half the shell's diameter. Some species bury their feces in a hole dug by the snout. Earliest fossil records date to the early Upper Cretaceous.

ABOVE | Dorsal view of the shell of *Xenophora pallidula*, a deep-water species from the Philippines. Attached shells double the effective diameter of the shell, deterring predators.

RIGHT | Ventral view of the shell of *Stellaria solaris*, an Indo-Pacific species. The spines raise the base of the shell above the surface of the sand.

REPRODUCTION
Sexes are separate. Fertilization is internal. Little is known about eggs or development, but protoconch morphology indicates that larvae are planktotrophic in all but one species that may have direct development.

LITTORINIMORPHA—CALYPTRAEOIDEA—CALYPTRAEIDAE
SLIPPER SHELLS, CUP AND SAUCER SHELLS, HAT SHELLS

Members of the family Calyptraeidae have adapted to a sedentary life on hard substrates that is more typical of epifaunal bivalves such as mussels and oysters. They are immobile and lift the edge of their shell slightly off the hard surface on which they have settled and use their large gills to create currents and filter particles and phytoplankton that they ingest. Their shells have few whorls, a very broad, rapidly expanding aperture, and approach limpet-like morphology to varying degrees. Shells of *Trochita* and *Sigapatella* may span several whorls and their columella remains coiled. Shells of *Crepidula* are cap-like with the columella forming a shelf separating the foot from the viscera. Shells of *Crucibulum* are conical, with a central apex and a cup-like columella.

All genera lack an operculum. The exterior of the shells may be smooth, or sculptured with

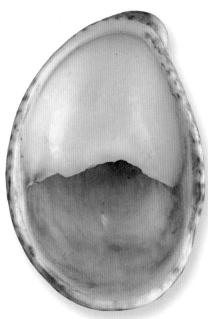

DISTRIBUTION
Temperate and tropical latitudes, but the family is also represented in polar regions.

DIVERSITY
Family includes 129 living species assigned to 11 genera within 1 subfamily.

HABITAT
Most species cling to hard substrates, which may include shells inhabited by crabs, in shallow-water marine and estuarine environments at depths ranging from intertidal to 200 ft (60 m).

SIZE
Species range in size from ¼–2½ in (6–60 mm).

DIET
Sedentary filter feeders.

OPPOSITE LEFT | Ventral view of the shell of *Crucibulum spinosum* with a cup-like columella to which the animal is attached.

OPPOSITE RIGHT | Ventral view of the shell of *Crepidula fornicata*, showing the broad shelf that separates the foot from the internal organs.

ABOVE | A tower of four specimens of *Crepidula fornicata*.

commarginal growth lines and axial ribs, some with scales or spines. Studies have shown that a limpet-like morphology developed early in calyptraeid evolution and that two lineages (*Trochita* and *Sigapatella*) have independently reverted to a coiled shell.

Animals have a muscular foot with which they hold the substrate and are capable of limited mobility. Adaptations for filter feeding include an elongated mantle cavity, an enlarged gill, and a food groove along which a mucus-bound strand of food particles is moved toward the mouth.

In some species, juveniles may use their radula to rasp food from the substrate until their gill has grown enough for filter feeding. Pelagic larvae may settle on the hulls of ships and their spawn have become invasive species as the ships travel well beyond the natural range of the species.

Earliest records of Calyptraeidae appear in the Lower Cretaceous (113–100 mya) with increases in geographic range and diversity during the late Eocene (49–34 mya).

REPRODUCTION
Protandrous hermaphrodites, changing from male to female as they grow. Some species of *Crepidula* form towers of multiple individuals. The largest and oldest at the bottom are females. As the individuals at the bottom die, some of the males above will change to females. Larvae settle at the top of the tower and begin adult life as males. Egg masses are brooded under the shell of the female and young are released as veliger larvae or as crawling young.

LITTORINIMORPHA—CAPULOIDEA—CAPULIDAE
CAP SHELLS

This family of morphologically diverse snails inhabits hard substrates at subtidal depths, primarily along continental shelves throughout the world. Shells range from tall to broadly conical, with a narrowly to widely open umbilicus, some with open coiling in later whorls.

Others have cap-like shells with very large apertures and few, rapidly expanding whorls. Species

BELOW | Two individuals of *Trichotropis cancellata* depositing eggs in capsules attached to a rock.

DISTRIBUTION
Global, mostly along continental shelves from the tropics to polar regions.

DIVERSITY
Family includes 139 living species assigned to 20 genera within 1 subfamily.

HABITAT
Epifaunal, living on hard substrates at sublittoral to bathyal depths. Some attach to the shells of other mollusks or to polychaete tubes.

SIZE
Species range in size from ½–3 in (12–75 mm).

DIET
Suspension feeders; some are opportunistic kleptoparasites, stealing food from filter-feeding invertebrates but do not feed on the tissues of the hosts.

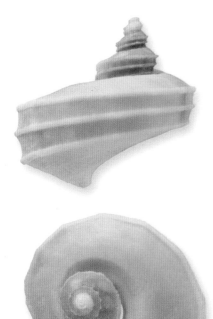
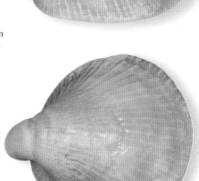

LEFT | *Separatista separatista*, a subtidal species from the southwestern Pacific in which the later whorls become detached.

RIGHT | *Capulus ungaricus*, a locally common offshore species from Iceland to the Mediterranean.

with coiled shells retain their operculum, which is lost in species with limpet-like or cap-shaped shells. The aperture may be round to rectangular in shape, with strong shoulders and short siphonal canals present in some genera. The outer lip is thin, the columella thickened but without plicae. The shell surface has spiral threads, cords, or ribs. Axial sculpture is either absent or limited to small ribs. The periostracum is thick and brown, and may have weak to strong hairlike projections along the shoulder and spiral cords.

REPRODUCTION
Protandric hermaphrodites. Sperm and eggs may be produced by the same individual during the transition from male to female. Fertilization is internal. Some species deposit eggs in capsules; others brood eggs within the mantle cavity, producing planktotrophic veliger larvae.

Animals have a characteristic pseudoproboscis, a long, tube-like extension of the mouth that is split along its dorsal surface and used to obtain food. Some species are obligate suspension feeders. Others are facultative kleptoparasites, stealing food particles from tube-dwelling polychaetes by extending their pseudoproboscis into openings of worm tubes. Some species live on the dorsal surfaces of scallops. They drill a hole in a scallop's shell over or near its mouth and insert the pseudoproboscis into the mantle to steal food that the scallop has filtered using its gills.

Another distinctive feature of Capulidae is their echinospira larvae, in which the larval shell is enveloped in a second shell that helps maintain buoyancy of the pelagic larvae. The family Capulidae dates to the Late Cretaceous (Campanian 83.6–72.1 mya).

LITTORINIMORPHA—CYPRAEOIDEA—CYPRAEIDAE
COWRIES

The shells of Cowries, with their distinctive egg shape, often elaborate color patterns, and glazed surface, have been used by humans since prehistoric times. They have functioned as money, jewelry, fertility symbols, badges of authority, and been incorporated in folk medicine and divination rituals for much of human history. Most Cowries live in relatively shallow depths along coral reefs and adjacent sandy areas throughout the tropics, feeding on algae or on sponges.

As juveniles, cypraeids produce smooth spiral shells with very wide apertures and thin outer lips. When the animal reaches adulthood, the direction of growth of the outer lip is abruptly deflected inward, with the last shell whorl completely enveloping the earlier whorls. Both the outer lip and the columellar area become greatly thickened, forming a narrow aperture that spans the length of the shell, with teeth along both inner and outer margins. The mantle completely envelopes the outer surface of the shell with the two lobes meeting along the dorsal midline. The shell no longer increases in length but becomes thickened as the mantle continues to add layers of shell, contributing to the often-complex color patterns on the dorsal surface of the shell as well as the glossy glaze. The surface of the mantle may have elaborate papillae that provide camouflage. The positions of the papillae often correspond to spots or pigmentation in the

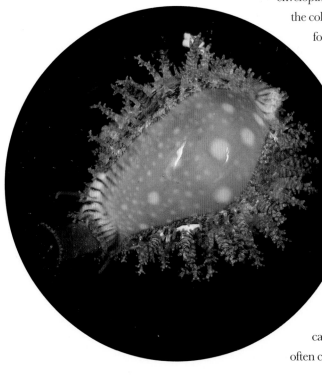

DISTRIBUTION
Tropical and subtropical seas worldwide, with greatest diversity in the tropical Indo-Pacific region.

DIVERSITY
Family includes approximately 200 living species assigned to 53 genera within 9 subfamilies.

HABITAT
Epifaunal, living on coral reefs and sandy bottoms at sublittoral depths, with some species occurring on outer reefs at depths of up to 2,100 ft (700 m). Most are nocturnal.

SIZE
Species range in size from ½–7½ in (12–190 mm).

LEFT | *Talparia talpa* off the coast of Hawai'i, crawling at night with mantle partially covering the shell.

BELOW | *Palmadusta ziczac* crawling along a coral reef.

shell. Neither a periostracum nor an operculum is present. The foot is broad and extensible. A broad siphon extends from the anterior margin of the shell. Cowries are nocturnal, remaining hidden during the day and emerging to feed at night.

Cypraeidae are represented in the fossil record since the Late Jurassic (Tithonian 152.1–145.0 mya).

OPPOSITE | *Perisserosa guttata guttata* with extended siphon and mantle as well as complex papillae partially enveloping the shell.

BELOW RIGHT | *Staphylaea staphylaea*, an Indo-Pacific species that lives in and under coral blocks. Unlike many Cowries, this species maintains a pustulose dorsal surface.

DIET
Primarily herbivores and omnivores. Some species feed on sponges.

REPRODUCTION
Sexes are separate. Males are usually smaller than females. Fertilization is internal. Eggs are deposited in capsules in gelatinous masses and hatch as veliger larvae. Females often remain with the eggs until they hatch.

LITTORINIMORPHA — CYPRAEOIDEA — OVULIDAE
EGG COWRIES, FALSE COWRIES

Species of the family Ovulidae are carnivorous snails that live on corals and sea fans, consuming their hosts' polyps and tissues. These ectoparasites are highly adapted to their habitat. Although some resemble Cowries to varying degrees, the shells of most genera are elongated and narrow with the aperture spanning the entire length of the shell and enveloping prior whorls. The outer lip turns inward at maturity and becomes thickened. It may or may not have denticles. The columella is smooth but may have a thickened node (funiculum) at its posterior margin. The shell surface is smooth and glossy and may have fine spiral threads on its surface. Colors range from white to brown, pink, or purple.

As in Cowries, the mantle completely envelopes the outer surface of the shell and may be smooth or have simple or complex papillae that are often brightly colored and patterned to mimic the color and polyp patterns of its host.

Ovulidae occur broadly throughout tropical and subtropical seas, with greatest diversity in the Indo-Pacific region. They range from intertidal depths to the deep sea, but their distribution is

LEFT | *Cyphoma gibbosum*, a tropical western Atlantic species, crawling on a soft coral.

DISTRIBUTION
Majority inhabit tropical and temperate latitudes worldwide, with greatest diversity in the Indo-Pacific region.

DIVERSITY
Family includes approximately 250 living species assigned to 39 genera within 4 subfamilies.

HABITAT
Ectoparasites of corals and sea fans that occur in shallow waters and the deep sea.

SIZE
Species range in size from ½–4½ in (12–115 mm).

DIET
Coral and sea fan tissues and polyps.

REPRODUCTION
Sexes are separate. Fertilization is internal. Females deposit eggs in capsules attached to the surface of their host corals, with larvae hatching as free-swimming veligers. Females of some species brood the capsules beneath their foot until they hatch.

ABOVE | *Diminovula stigma* crawling on a soft coral upon which it feeds in the northwestern Pacific.

RIGHT | Dorsal and apical views of the shell of *Aclyvolva lanceolata* from the tropical western Pacific.

BELOW | Dorsal and apical views of the shell of *Volva habei*, a Japanese species reaching 2½ in (63.5 mm) in length.

BELOW | Dorsal view of the shell of *Phenecovolva nectarea*, a species with a wide range throughout the Indo-West Pacific.

limited by the availability of suitable host species. Some surveys have indicated that nearly half of the encountered species of soft corals did not serve as hosts for ovulids, possibly because they produce secondary metabolites that protect against predation.

Oldest fossil records of Ovulidae appear in Late Cretaceous deposits (Cenomanian 100.5–93.9 mya).

LITTORINIMORPHA—NATICOIDEA—NATICIDAE
MOON SNAILS

Predatory Moon Snails live in sand and mud bottoms from the tropics to the poles and from intertidal to abyssal depths.

Most Moon Snail shells are globular to ovate in shape, with a low spire with few whorls and a very large inflated final whorl. The aperture is very large and ovoid in outline, without a siphonal canal. Some have an open umbilicus; others fill the umbilicus with a callus. The shells may be smooth, with fine growth lines, or glossy. Opercula have an eccentric nucleus and are entirely corneous in some subfamilies and calcified and elaborately sculptured in others.

The animal inflates its body with seawater to several times the volume of the shell to expand the foot and lateral flaps that can partially or completely envelope the shell and a large head shield that is reflected over the head and tentacles. This allows the animal to plow through the sand or mud to hunt for infaunal prey. However, the animal can expel the water and retract completely into the shell when threatened.

When a Moon Snail encounters its prey, most often a bivalve, it almost completely envelops it with its foot and drills a hole by alternating applications of an enzymatic secretion from its accessory boring organ, located below the mouth at the anterior of the proboscis, and scraping partially dissolved shell with its radula until the shell is penetrated. It then inserts its proboscis through the hole to consume the prey. The holes produced by Moon Snails are easily recognized by their circular shape and broadly beveled edges. They can be distinguished from holes produced by muricids, which are narrower in diameter and have straight sides.

Moon Snails have an extensive fossil record dating to the Late Triassic (227 mya).

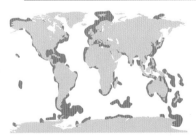

DISTRIBUTION
Global, with greatest diversity in tropical regions, but extending to both polar regions.

DIVERSITY
Family includes approximately 370 living species assigned to 39 genera within 4 subfamilies.

HABITAT
Semi-infaunal, in sand and mud substrates, from intertidal to abyssal depths.

SIZE
Species range in size from ⅛–6½ in (3–165 mm).

DIET
Infaunal bivalves, scaphopods, and other gastropods. One species preys on crabs.

RIGHT | *Euspira heros* off the coast of Massachussetts. The foot and head shield nearly envelope the shell.

BELOW LEFT | A venerid bivalve with a characteristic borehole produced by a naticid.

BELOW RIGHT | *Naticarius orientalis*, a shallow-water species from the tropical Indo-Pacific.

OPPOSITE | *Neverita lewisii*, the largest living naticid species, occurs along the coast of western North America from British Columbia, Canada to Baja California, Mexico.

REPRODUCTION

Sexes are separate. Fertilization is internal. Females deposit their eggs in capsules embedded in a "sand collar" composed of sand grains cemented by mucus, with the inner edge conforming to the aperture of the shell. Eggs hatch as planktonic veligers or develop directly into crawling juveniles, depending on species.

LITTORINIMORPHA—TONNOIDEA—CASSIDAE
HELMET SHELLS, BONNET SHELLS

Bonnet Shells, or Helmet Shells, are large, predatory snails adapted to feed on echinoderms, primarily sea urchins. They burrow into the sand during the day and emerge at night to feed.

Species vary considerably in size. Shells tend to be globose or ovoid in shape, with a low spire and a final whorl that spans most of the shell. Growth generally occurs rapidly in increments of 2/3 whorl (270 degrees), terminating in the secretion of a thickened columellar callus and an enlarged outer lip, often with teeth along its edge, that define a flattened ventral surface of the shell, the columellar shield. The positions of earlier columellar shields are evident along the spire. The aperture is ovate to elongate, with an anterior, dorsally deflected siphonal canal. The operculum is corneous, thin, and similar in shape to the aperture, but does not completely fill it. The shell surface has spiral and axial sculpture that vary in prominence, and may be uniformly colored or with complex patterns, often most conspicuous on the varices. The head is large with long tentacles with eyes at their bases.

LEFT | *Cassis cornuta* prepares to feed on a sea urchin off Hawai'i.

DISTRIBUTION
Global, with most species inhabiting tropical and temperate regions.

DIVERSITY
Family includes 97 living species assigned to 12 genera within 2 subfamilies.

HABITAT
Most live in sandy substrates at depths ranging from subtidal to about 330 ft (100 m).

SIZE
Species range in size from 1½–16½ in (38–410 mm). Most are 2–5 in (50–125 mm).

DIET
Carnivores that feed on echinoderms, primarily sea urchins.

The animal crawls onto its prey with its large foot and appears resistant to the toxins in the spines of the urchin. It then drills a hole through the urchin's shell using a secretion from its salivary glands that includes sulfuric acid and a neurotoxin, then inserts its long proboscis and consumes the tissues.

Helmet shells, especially the larger and heavier species, have long been used to carve cameos.

The fossil record of the family Cassidae extends from the Late Cretaceous (Coniacian 89.8–86.3 mya).

LEFT | Dorsal and apical views of the shell of *Cassis flammea*, a common shallow-water species from the tropical Western Atlantic.

LEFT | Dorsal and apical views of the shell of Scotch Bonnet, or *Semicassis granulata*, the official state shell of North Carolina.

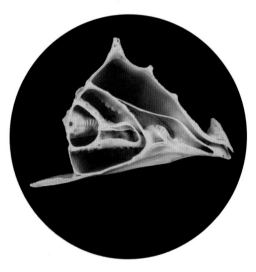

ABOVE | The shell of *Cassis cornuta* sectioned near its coiling axis.

REPRODUCTION
Sexes are separate. Males are smaller than females in some species. Fertilization is internal. Females deposit eggs in proteinaceous capsules, which form tower-like structures. In some species, multiple females may converge to produce a communal egg mass. Eggs hatch as veliger larvae.

LITTORINIMORPHA—TONNOIDEA—CYMATIIDAE
TRITONS

Cymatiidae are extremely diverse in morphology and ecology of included species. Most are moderately large inhabitants of tropical and temperate continental shelf habitats that include rocky shores, coral reefs, and sandy bottoms, but some genera occur at bathyal depths (to 3,280 ft / 1,000 m) along upper continental slopes.

Cymatiidae have solid shells, triangular to pyriform in outline, with moderately tall spires and a siphonal canal that may be short or long and irregularly curved. Many have pronounced axial ribs and spiral cords that form nodes at their intersections and two or fewer varices per whorl that are usually offset by 2/3 whorl, but in some species align with those of the previous whorls to form wing-like

DISTRIBUTION
Most occur in the tropics but some genera extend to temperate and polar regions.

DIVERSITY
Family includes approximately 150 living species assigned to 23 genera within 1 subfamily.

HABITAT
Most inhabit rocky shores, coral reefs, and sandy bottoms from intertidal to outer continental shelf depths (0–656 ft / 0–200 m), but several occur at bathyal depths of up to 3,280 ft (1,000 m).

SIZE
Species range in size from 1¼–10 in (32–254 mm).

structures. The aperture is usually large and ovate, and often has pronounced lirae along the columella and outer lip. The periostracum is usually well-developed, often with long, hair-like bristles. The operculum is thick and corneous. A large protoconch comprised of multiple whorls is present in most genera.

The head, broad foot, and mantle of many species have elaborate color patterns, often of circular spots with well-defined edges that differ in color. The head has long tentacles with well-developed eyes at mid-length. Cymatiidae are predators that feed on a wide variety of invertebrates, including other mollusks, a variety of echinoderms including sea cucumbers, sea urchins, and starfish, as well as ascidians and tube worms. Some species have a narrower range of prey than others.

Although several species undergo direct development and hatch as crawling young, most have very wide geographic ranges because they hatch as pelagic larvae and have unusually long planktonic larval stages, allowing them to be widely dispersed by ocean currents. Females of some species remain with the egg capsules they deposit until they hatch, which may take three weeks to three months.

The fossil record extends to the Early Cretaceous (Barremian—125–129.4 mya).

ABOVE | Dorsal view of the shell of *Gyrineum perca*, commonly known as the Maple Leaf Triton, from the Western Central Pacific.

RIGHT | Dorsal view of the shell of *Cymatium femorale*, a wide-ranging species from the tropical Western Atlantic.

OPPOSITE | *Monoplex parthenopeus* emerging from its shell, which is covered by a thick, hirsute periostracum.

DIET
Epifaunal carnivores, feeding on other mollusks, echinoderms including sea cucumbers (holothurians), ascidians, tube worms, and other invertebrates.

REPRODUCTION
Sexes are separate. Fertilization is internal. Females deposit hemispherical egg masses, each containing nearly 200 individual egg capsules, with thousands of eggs per capsule. Eggs hatch as pelagic veliger larvae.

LITTORINIMORPHA—TONNOIDEA—TONNIDAE
TUN SHELLS

The name Tun Shells is due to their resemblance to a "tun," a large cask or barrel used to store liquids such as beer and wine. These predatory snails inhabit sandy bottoms in shallow to moderately deep water, mostly at tropical but also at temperate latitudes. During the day, they burrow into the sand with only the tips of their siphons exposed. Tun Shells emerge at night to prey primarily on echinoderms such as sea cucumbers, but also on bivalves, crustaceans, and fish.

Tun Shells have large, thin, inflated, and nearly spherical shells, with a short, broad spire and a very large body whorl. The aperture is broadly ovate with a short anterior siphonal canal in the form of a deflected U-shaped notch. Sculpture is of

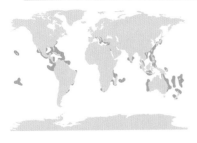

DISTRIBUTION
Tropical and temperate seas worldwide.

DIVERSITY
Family includes 35 living species assigned to 3 genera within 1 subfamily.

HABITAT
Sandy substrates from subtidal to bathyal depths.

SIZE
Species range in size from 2–12 in (50–300 mm).

DIET
Feed at night, primarily on sea cucumbers (holothurians) but also on other invertebrates and fish.

REPRODUCTION
Sexes are separate. Fertilization is internal. Females deposit eggs in broad gelatinous ribbons. Eggs hatch as larvae that spend 2–8 months in the plankton before settling to the bottom.

alternating spiral ridges separated by furrows of similar width, producing a corrugated appearance. The periostracum is thin and straw-colored. In the genus *Tonna*, the outer lip remains thin, or becomes thickened and forms denticles, as it does in the genus *Eudolium*. The absence of varices indicates that the shell is no longer growing once the outer lip is thickened. In species of *Malea*, the outer lip turns inward before flaring outwardly and developing large denticles, while the columella develops thick folds and lirae along its anterior and posterior margins with a smooth, U-shaped notch at its mid-length. An operculum is present in larvae, but lost in adults.

The head is broad and has elongated tentacles with eyes along the outer edges of their thickened bases. The foot is very large, thin, broad anteriorly with a distinct propodium, and tapered posteriorly. The siphon is long and tubular. The proboscis is relatively short, wide, and flattened at its end. Large salivary glands produce a secretion that contains sulfuric acid used to paralyze its prey, which is then enveloped by the broad proboscis and swallowed whole.

Origins of the family have been traced to the Late Cretaceous (Maastrichtian 72.1–66 mya).

OPPOSITE | *Tonna canaliculata* crawling along a rubble bottom in the Philippines.

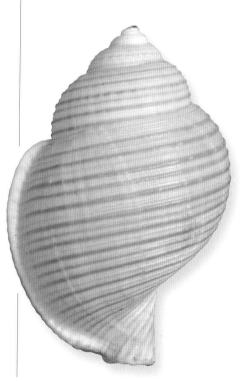

BELOW | *Eudolium bairdii* is a species with a circumtropical distribution mostly at bathyal depths throughout the Atlantic, Pacific, and Indian Oceans.

LITTORINIMORPHA — PTEROTRACHEOIDEA — CARINARIIDAE
HETEROPODS

Like the related families Atlantidae and Pterotracheidae (also components of the superfamily Pterotracheoidea), Carinariidae are adapted to an entirely pelagic environment, mostly near the surface, rarely below 650 ft (200 m) depth.

All carinariids are active, visual predators that feed on zooplankton and generally swallow their prey whole. They have an elongated, cylindrical body and a greatly reduced, delicate shell. The body is gelatinous to aid in buoyancy and is nearly transparent to deter predators. It consists of a head region with a large proboscis containing a muscular buccal mass with a radula and is flanked by short tentacles with large eyes at their bases. All the internal organs are concentrated in a stalked visceral mass situated on the dorsal posterior region of a broad trunk. The visceral mass together with the mantle cavity organs are covered by the shell. Posterior to the trunk is a tapering tail with

BELOW | *Carinaria lamarckii*, swimming at depths 243–617 ft (74–188 m) above the Mid-Atlantic Ridge in the North Atlantic Ocean.

DISTRIBUTION
Global, in the pelagic zone of tropical and temperate seas.

DIVERSITY
Family includes 9 living species assigned to 3 genera within 1 subfamily.

HABITAT
Epipelagic, occurring primarily as part of plankton communities dominated by salps, chaetognaths, and jellyfish.

SIZE
Species range in size from 1½–26 in (40–660 mm). Corresponding shell length ½–3 in (10–75 mm).

DIET
Feed primarily on salps (Tunicata) but may also include arrow worms (Chaetognatha), pelagic nudibranchs (Gastropoda), copepods, and euphasid shrimp as well as small fish and pelagic larvae.

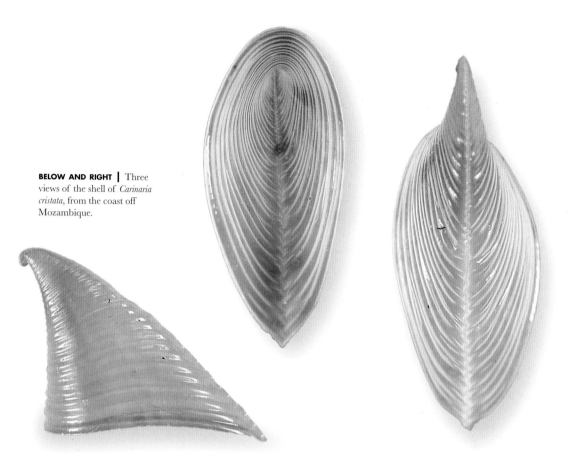

BELOW AND RIGHT | Three views of the shell of *Carinaria cristata*, from the coast off Mozambique.

a ventral keel opposite the visceral mass. The animal swims by undulating its body and flapping a large fin derived from the foot, with the fin directed toward the surface and the shell downward.

The shell is composed of aragonite and is very thin, fragile, and nearly transparent. The larval shell (protoconch) is tightly coiled, but after metamorphosis, the aperture expands rapidly, resulting in a laterally compressed cap-like shell with a triangular profile. The shell surface has broad, commarginal waves and a narrow dorsal keel.

The earliest record of the family Carinariidae has been reported from the Uppermost Lower Jurassic (Toarcian 174–182 mya) of southern Germany.

REPRODUCTION
Sexes are separate and there is sexual dimorphism. Males have a tri-lobed penis and produce spermatophores. Fertilization is internal. Females deposit eggs in a floating mucus string. Larvae hatch as planktotrophic veliger larvae.

CAENOGASTROPODA—EPITONIOIDEA—EPITONIIDAE

WENTLETRAPS

The family Epitoniidae is an ancient lineage that adapted to a specialized diet of Cnidaria early in its evolution.

Wentletraps is a term derived from the German "wendeltreppe," meaning spiral staircase. During the 17th and 18th centuries, the Precious Wentletrap (*Epitonium scalare*) was the most desired and valuable shell among European collectors. It has been rumored that these shells were so rare and valuable that forgeries were made from rice paste.

Shells of most species are white, very high-spired, with convex whorls that may be tightly wound, loosely coiled, or rarely disjunct, connected only by their axial ribs. Axial sculpture is prominent, with regularly spaced ribs or costae, with varices present in some taxa. Shell surfaces range from porcelaneous to chalky among genera. Some species may be umbilicate. The aperture is round, as is the tightly fitting operculum, which is chitinous and typically black or yellow.

Anatomical adaptations for feeding on cnidarians include a large head, broad snout, a characteristic ptenoglossate radula with fang-like teeth, large jaws, and cuticularized stylets that inject salivary secretions. The hypobranchial gland produces a purple secretion that may have anesthetic properties on the prey.

Several genera (*Recluzia* and *Janthina*) are pelagic, floating along the surface of tropical seas attached to a raft they produce by trapping bubbles of air in mucus. These genera have very thin, smooth, globular shells and feed on cnidarians that float on the ocean surface.

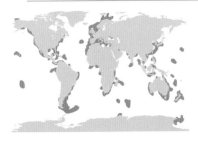

DISTRIBUTION
Global, at depths from intertidal to abyssal (over 13,000 ft / 4,000 m).

DIVERSITY
Family includes approximately 630 living species assigned to 49 genera within 1 subfamily.

HABITAT
Most genera are benthic, burrowing in sandy bottoms near sea anemones, or on or near rocks and corals. Some species of *Janthina* are abundant and may form large shoals many miles in diameter.

SIZE
Species range in size from ⅛–5 in (3–120 mm).

DIET
Predators or ectoparasites of cnidarians. Benthic species feed on sea anemones and corals while pelagic species feed on floating sea anemones (*Minyas* sp.) or siphonophores.

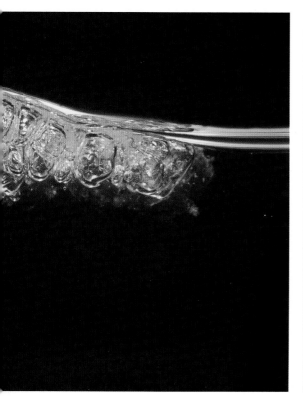

Earliest fossil records of Epitoniidae are from the Triassic of Europe and Indonesia, while the pelagic genera first appear in Miocene deposits of Australia and Japan.

REPRODUCTION
Protandrous hermaphrodites; males at first, changing to females as they grow. Fertilization is internal by transfer of spermatozeugmata. In benthic species, eggs are deposited in sand-covered capsules connected by a mucus strand attached hard substrates. The pelagic species attach egg capsules to the bottom of their floats. Some are ovoviviparous, retaining the eggs within their body until they hatch.

OPPOSITE | Animal of *Janthina janthina* hangs below its floating raft of mucus-coated bubbles.

ABOVE RIGHT | *Epidendrium aureum*, proboscis extended, feeding on a coral polyp. Recently deposited egg capsules are to its left. Found off New South Wales, Australia.

LEFT | Dorsal view of the shell of the Precious Wentletrap (*Epitonium scalare*). The whorls are disjunct, connected only along the axial ribs.

LITTORINIMORPHA—VANIKOROIDEA—EULIMIDAE
EULIMID SNAILS

The family Eulimidae comprises one of the largest and most diverse lineages of parasitic gastropods. All eulimids are parasitic on echinoderms, but different genera and species are associated with each of the five classes of echinoderms, including sea urchins, sea cucumbers, starfish, brittle stars, and sea lilies. Many are free-living ectoparasites that penetrate the skin of their hosts with their proboscis until they find a suitable organ and then feed on the body fluids of the host. Others are endoparasites, burrowing into and living within the tissues of the host on which they feed. Still others live within galls they produce in the tissues of their host.

Shells of most ectoparasitic eulimids have a high conical spire and are smooth, solid, glossy, and white, translucent, or with colored patterns. Others have surface sculpture of prominent spiral bands, nodules, or axial ribs. The aperture is small and elliptical, and the operculum is thin and corneous. However, some ectoparasitic species have limpet-like shells. Shells of endoparasitic eulimids and species that

BELOW | *Thyca crystallina*, feeding on a blue starfish (*Linckia laevigata*).

DISTRIBUTION
Global, and are parasitic on echinoderms in all oceans, ranging from intertidal to abyssal depths.

DIVERSITY
Family includes approximately 1,000 living species assigned to 103 genera within 1 subfamily.

HABITAT
Eulimidae are parasitic on echinoderms. Some are ectoparasites; others are endoparasites, and some form galls within the tissues of their hosts.

SIZE
Species range in size from $^3/_{64}$–1 ¼ in (1–30 mm).

DIET
Parasitic on Echinodermata. Different species and genera are specific to each of the five classes of echinoderms.

ABOVE | Two specimens of *Annulobalcis yamamotoi* at the base of a crinoid on which they feed. Found off New South Wales, Australia.

LEFT | *Niso splendidula* from off the Pacific coast of Panama.

REPRODUCTION
Some species have separate sexes; others are either simultaneous or protandric hermaphrodites. Most species produce egg capsules with multiple eggs that are attached to the host or to the substrate. Some species have planktotrophic; others lecithotrophic larval development.

form galls tend to be fragile, thin, often rounded, and more varied in shape. The shell may be absent in some specialized endoparasites.

Animals also vary widely, and some are brightly colored. Most have a long and slender foot that secretes a mucous thread with which the animal attaches to the host. The head is narrow, and tentacles are long with eyes at their bases. All eulimids lack a radula and jaw but have a long proboscis.

The oldest eulimid fossils date from the Late Cretaceous (Cenonanian–Maastrichtian 100.5–66 mya), and are much younger than their echinoderm hosts, which date to the Paleozoic (more than 250 mya).

NEOGASTROPODA—CANCELLARIOIDEA—CANCELLARIIDAE
NUTMEG SHELLS

The family Cancellariidae is the most basal offshoot of the Neogastropod radiation that dates to the Early Cretaceous and is the sister group to the remaining Neogastropoda. The family has diversified during the Cenozoic.

Many shells are globular- to barrel-shaped, but some lineages tend to be fusiform; others strongly shouldered or with open coiling. Most species have prominent axial and spiral sculpture resulting in a reticulate or cancellate surface for which the family is named. Some cancellariids are said to resemble nutmeg seeds, hence the common name Nutmeg Shells.

The aperture is wide and contains two or three columellar folds that become expanded in apposition to correspondingly enlarged teeth on the outer lip, usually every 1/3 to 2/3 whorl, forming "internal varices" that reinforce the shell against crushing predation. Members of the subfamily Admetinae that live in polar regions and the deep sea tend to be thin-shelled and lack significant surface sculpture, columellar folds, or apertural teeth. Opercula are present in larvae but absent in adults.

Cancellariidae are distinguished by their unique anterior alimentary system adapted for suctorial feeding. They have a long, extendable proboscis that contains a distinctive radula with a single row of very long, ribbon-like teeth with complex, interlocking cusps at the ends, a highly modified jaw that forms a tubular structure through which the teeth pass and which envelopes the buccal mass, and a simplified alimentary system. Some species of Admetinae lack a radula but

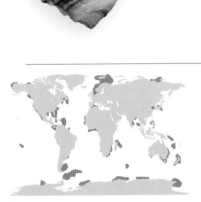

DISTRIBUTION
Global, primarily at continental shelf depths but some species inhabit depths of up to 3,280 ft (1,000 m).

DIVERSITY
Family includes approximately 350 living species assigned to 52 genera within 3 subfamilies.

HABITAT
Sand and mud bottoms.

SIZE
Species range in size from ½–3 in (13–76 mm).

DIET
The blood of fish as well as bodily fluids of invertebrates and contents of egg capsules.

LEFT | *Trigonostoma milleri*, a species with open coiling, from subtidal depths off the tropical eastern Pacific.

share the other features of the alimentary system. Recent observations have documented *Cancellaria cooperii* feeding on the blood of sleeping electric rays, giving rise to the common name Vampire Snails. Other species of cancellariids have been reported to feed on body fluids of bivalves, gastropods, and contents of egg capsules.

BELOW | *Scalptia contabulata* on a subtidal sand bottom in the Philippines.

OPPOSITE | Dorsal and apertural views of the shell of *Cancellaria reticulata*, a shallow-water species from the tropical and temperate western Atlantic.

REPRODUCTION

Sexes are separate and fertilization is internal. Females deposit fertilized eggs in spatulate capsules with a preformed hatching aperture and supported by a stalk that may be long in sand-dwelling species and shorter in species dwelling on or near rocks. The number of eggs per capsule ranges from 5 to 5,000, and mode of development—ranging from planktotrophic, lecithotrophic, or direct development—varies with species.

NEOGASTROPODA—VOLUTOIDEA—VOLUTIDAE
VOLUTES, BALER SHELLS

The Volutidae is among the more basal families within the Neogastropoda, an order of carnivorous gastropods that originated and radiated rapidly during the Upper Cretaceous and Paleogene.

Shell shapes vary from elongate and fusiform to broadly ovate, and sometimes involute. Some are uniformly colored and covered with a thick brown periostracum; others are colorful and exhibit elaborate patterns with either a thin periostracum or a glazed surface. The aperture is elongated and may be rounded or strongly shouldered, with knobs or spines along the shoulder. A short anterior canal is present with a broad, rounded siphonal notch. The

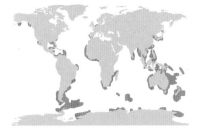

DISTRIBUTION
Global, with the highest diversity being in Australia and adjacent waters.

DIVERSITY
Family includes approximately 250 living species assigned to 47 genera within 8 subfamilies.

HABITAT
Most species live on shallow-water sand bottoms throughout tropical and temperate latitudes; others are adapted to bathyal and abyssal depths and polar latitudes.

SIZE
Species range in size from 5/16 in (8 mm) to over 2 ft (0.6 m). Most are 2–8 in (50–200 mm).

DIET
Carnivores that feed primarily on other mollusks, which they envelope with their foot.

columella is smooth and usually has three to five prominent plaits, sometimes with additional weaker plaits. Shells of some large species (called baler shells) were used to bail canoes in the tropical Indo-Pacific.

Animals of tropical species tend to have bold color patterns or spots often corresponding to color patterns of the shell. The foot is large, broad, and flattened. A small, corneous operculum occurs in several subfamilies, but is absent in most volutes. The head is wide and flattened with lateral lobes bearing tentacles with eyes at or slightly behind their bases.

The siphon is long, broad, and flanked by one or two (varying with subfamily) lobes at its base. In some genera, the mantle may envelope the shell, producing a smooth glaze on its surface.

Volutes have an extensible proboscis. Several species retain three teeth per row on their radular ribbon, but most have only a single tooth per row, which varies considerably among taxa. Some have wide teeth with multiple, comb-like cusps; others have wishbone-shaped teeth with only a single cusp.

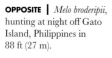

LEFT | Apical view of the shell of *Cymbiola imperialis*, a shallow-water species from the southern Philippines.

BELOW LEFT | *Voluta musica*, from shallow-water sandy bottoms in the Caribbean.

BELOW RIGHT | *Scaphella junonia* occurs in offshore sandy bottoms in the US, from North Carolina to the Florida Keys and in the Gulf of Mexico.

OPPOSITE | *Melo broderipii*, hunting at night off Gato Island, Philippines in 88 ft (27 m).

REPRODUCTION
Sexes are separate. Fertilization is internal. Females deposit eggs in proteinaceous capsules formed by a gland in the sole of their foot and attached to hard substrates. Metamorphosis occurs within the egg capsules and juveniles emerge as crawling young.

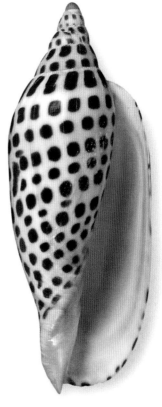

NEOGASTROPODA—BUCCINOIDEA—BUCCINIDAE
WHELKS, TRUE WHELKS

Buccinidae comprise the most geographically widespread and ecologically diverse family of neogastropods, ranging from the tropics to polar regions of all oceans, and inhabiting depths ranging from the intertidal zone to ocean trenches, including hydrothermal vents. Greatest diversity occurs on hard substrates in sublittoral to bathyal depths.

Shells of most species have a conical spire with whorls that may have an angular shoulder or be smoothly rounded. The aperture is ovate with an outer lip that is generally smooth but may form a corrugated edge in some taxa. The columella is smooth and rounded. Some species have a siphonal canal of moderate length; others a broad siphonal notch. Surface sculpture ranges from smooth and evenly rounded to reinforced by axial ribs or spiral cords that vary in strength. The periostracum is moderately thick and brown. The operculum is corneous, thin, and oval to claw-shaped, with its nucleus either central or terminal.

The animal has a foot that is broad anteriorly and rounded posteriorly, a long, narrow head with two tentacles and eyes at their bases, and a long siphon that is used to detect and locate prey. The proboscis can be extended to nearly twice the length of the body. The salivary glands of buccinids were

BELOW | *Buccinum undatum* is a common cold-water whelk on both sides of the Atlantic.

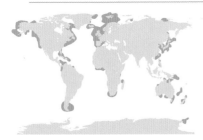

DISTRIBUTION
Global, from intertidal to hadal depths, including hydrothermal vents.

DIVERSITY
Family includes approximately 1,000 living species assigned to 29 genera within 8 subfamilies.

HABITAT
Most species are epifaunal, living on solid bottoms, but many are infaunal or semi-infaunal, living on sand and mud substrates.

SIZE
Species range in size from 1–4 in (25–100mm) but some can reach 10 in (250 mm).

DIET
Predators, feeding primarily on other invertebrates, including other gastropods, bivalves, crustaceans, worms, and echinoderms, but are also necrophagous carnivores, scavenging dead vertebrates and invertebrates.

LEFT | *Neptunea tabulata* inhabits offshore sand and mud bottoms from British Columbia to southern California.

ABOVE | *Neptunea antiqua*, a female depositing eggs in leathery capsules that form a cluster of 30 or more capsules.

found to contain high levels of tetramine, a neurotoxin that may be used to subdue its prey.

Life spans of several Buccinidae species have been determined to range from 11 to 18 years. Buccinidae, together with the Fasciolariidae, are the two oldest families of Neogastropoda, both first appearing in the fossil record from the Early Cretaceous (Valanginian 139.8–132.9 mya).

REPRODUCTION
Sexes are separate. Fertilization is internal. Females deposit eggs in leathery capsules that may include nurse eggs. Some taxa hatch as crawling juveniles; others have a planktonic larval stage.

NEOGASTROPODA—BUCCINOIDEA—BUSYCONIDAE
LIGHTNING WHELKS

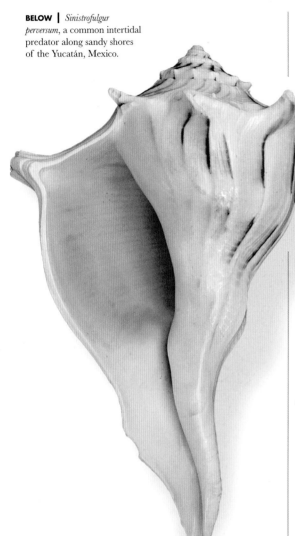

BELOW | *Sinistrofulgur perversum*, a common intertidal predator along sandy shores of the Yucatán, Mexico.

Since its origins during the Cretaceous (70.6 mya) the Busyconidae (recently elevated to the rank of family) has been endemic to shallow coastal waters of eastern North America and the Gulf of Mexico. During the Cenozoic, it diversified into the subfamilies Busyconinae and Busycotypinae.

Busyconidae species are the largest gastropods throughout much of their range. Shells are pyriform, with a moderately low spire, a pronounced shoulder, and a broad tapering siphonal canal. The body whorl is inflated, and the columella is long and smooth. Axial sculpture may consist of open spines or tubercles along the shoulder corresponding to increments of growth or be reduced or absent in some smaller species. The protoconch consists of two whorls, the first globose. Many of the species have irregular longitudinal stripes along the shell, especially as juveniles, which led to its common name of Lightning Whelk. Most species are dextrally coiled (aperture on the right side of the coiling axis), but the genus *Sinistrofulgur*, which has the broadest geographic range, is sinistrally coiled (aperture on the left side of the coiling axis). The operculum is thick, corneous, has a terminal nucleus, and fills most of the aperture.

DISTRIBUTION
Cape Cod, Massachusetts southward around Florida, along the coast of the Gulf of Mexico to the northern shore of the Yucatán Peninsula.

DIVERSITY
Family includes approximately 16 living species assigned to 6 genera within 2 subfamilies.

HABITAT
Burrows in sand and mud bottoms along the shore and in lower estuaries from intertidal depths to the outer continental shelf.

SIZE
Species range in size from 4–18 in (100–457 mm).

DIET
Infaunal predators feeding on bivalves, other invertebrates, and occasionally on carrion.

LEFT | A strand of egg capsules of *Sinistrofulgur sinistrum*. Individual capsules are attached to a central strand. The initial five to eight capsules, which do not contain eggs, are buried in the sand to anchor the strand, The remaining capsules remain above the sand.

BELOW | Animal of *Busycotypus canaliculatus* in the process of reorienting itself prior to burrowing into the sand.

The animal has a broad, crawling foot, a thick, muscular proboscis, and a radula with three teeth per row. The central tooth has three cusps in species of Busycotypinae, and four to seven cusps in species of Busyconinae.

Females are much larger than males and may mate with multiple males before producing a strand of over one hundred egg capsules anchored in the sand. The strands are coiled in the direction opposite of the female that produced them. Dextral species produce sinistrally coiled egg stands and vice versa.

REPRODUCTION

Sexes are separate. Fertilization is internal. Females may mate with multiple males prior to producing a strand of egg capsules with thirty or more eggs per capsule. Juveniles hatch as crawling young and do not have a pelagic stage.

NEOGASTROPODA—BUCCINOIDEA—FASCIOLARIIDAE
HORSE CONCH, TULIP SHELLS, SPINDLE SHELLS

Fasciolariidae are one of the oldest lineages within the order Neogastropoda, which first appeared during the Early Cretaceous (Valengian 139.8–132.9 mya). These large predatory snails now inhabit a broad range of habitats throughout the world. Most live on subtidal sandy bottoms throughout the tropics, but others inhabit deeper waters and higher latitudes.

Shells are fusiform, with high spires that may have rounded whorls or be strongly shouldered but lack varices. Siphonal canals are present and may be short to very long and axially oriented. The aperture may be ovate to hemi-elliptical and have an outer lip that is smooth or lined with multiple spiral lirae. The columella has a siphonal fold and may be smooth or have up to four plicae. The surface may be smooth or have spiral cords and axial ribs that are most pronounced along the shoulder. Shell color ranges from white to brown and may have darker axial or spiral bands. The periostracum is thick and tan to brown in color. The claw-shaped operculum is thick and corneous, with a terminal nucleus.

The head and foot of the animal tends to be brightly colored, ranging from red to orange, yellow or black, often with fine lighter or darker spots or speckles. The head is small, with short tentacles with

DISTRIBUTION
Cosmopolitan, with highest diversity at tropical and temperate latitudes.

DIVERSITY
Family includes approximately 540 living species assigned to 66 genera within 3 subfamilies.

HABITAT
Sand and mud bottoms and in seagrass beds, from the intertidal zone to depths of 2,600 ft (800 m).

SIZE
Species range in size from ¾–24 in (20–600 mm).

DIET
Carnivores; larger species feed primarily on other gastropods and bivalves, while smaller species feed on polychaete worms and barnacles.

LEFT | *Cinctura hunteria* depositing eggs in urn-shaped capsules along the west coast of Florida.

BELOW | *Fusinus crassiplicatus* inhabits sandy bottoms at continental slope depths throughout the tropical Indo-West Pacific.

eyes at their bases. The proboscis is extensible and very long. Larger species prey on other gastropods, which they envelop with their foot. Smaller species tend to feed on worms, and some have specialized to feed on barnacles.

OPPOSITE | *Triplofusus giganteus*, the largest gastropod in the Atlantic Ocean and the world's second-largest living species of gastropod, hunting along a sandy subtidal bottom.

REPRODUCTION
Sexes are separate. Fertilization is internal. Females deposit eggs in urn-shaped, leathery capsules attached to hard substrates. Juveniles hatch as crawling young, having been nourished within the egg capsules by consuming nurse eggs.

LEFT | *Fasciolaria tulipa* is a common predator on seagrass beds in bays and estuaries throughout the tropical Western Atlantic.

NEOGASTROPODA—BUCCINOIDEA—NASSARIIDAE
MUD SNAILS, NASSA SNAILS

Nassariidae are closely related to Buccinidae, often occurring in densities of hundreds of individuals per square yard. They burrow into the substrate with only their siphons above the surface of the sand and emerge to forage for food as the tide goes out.

They have shells with a high spire, a rounded anterior, and an ovate aperture with either a short, dorsally deflected siphonal canal or a broad siphonal notch anteriorly. The outer lip may be smooth or lyrate. The columella is smooth and may be thickened to form a broad callus or shield over the ventral surface of the shell. An exhalant channel at the posterior margin of the aperture is defined by a denticle. The shell surface may be smooth, glazed, or sculptured with axial ribs, nodules, and/or spiral cords. Shell color varies from white to dark brown and may have regular or irregular bands or blotches. The operculum is small, corneous, claw-like, and may have a serrated edge.

Animals have a broad, elongated foot that tapers posteriorly and has two metapodial tentacles at its posterior margin. The head has two long tentacles with eyes at their bases. When extended, the siphon is as long as the shell and is used to detect and locate food. The extensible proboscis is very long and used to access food in tight areas.

DISTRIBUTION
Tropical and temperate latitudes, but several species occur in polar regions.

DIVERSITY
Family includes approximately 600 living species assigned to 30 genera within 7 subfamilies.

HABITAT
Sandy and muddy bottoms in shallow, nearshore habitats and estuaries. Some occur on reefs and offshore in depths of several hundred feet.

SIZE
Species range in size from ¼–3 in (6–76 mm).

DIET
Carnivores and active scavengers that feed on carrion. Some may become facultative herbivores.

LEFT | *Tritia reticulata*, on seaweed, off the coast of Brittany, France.

BELOW | Dorsal and apical views of *Nassarius papillosus*, a large species that occurs throughout the tropical Indo-Pacific region.

Two genera have adapted to freshwater habitats in southeastern Asia, and are sold in the aquarium trade as Assassin Snails. Mud Snails appear to have originated during the Early Cretaceous (Aptian 125–113 mya) and diversified rapidly during the Cenozoic.

OPPOSITE | Egg capsules of *Tritia reticulata*, on seaweed, off the coast of Brittany, France. When the larva are ready to hatch, the region at the top of each egg capsule will dissolve and allow them to exit the capsule.

RIGHT | Dorsal and apical views of the shell of *Nassarius variciferus*, showing that this species undergoes episodic growth, each increment terminating in a white varix.

REPRODUCTION
Sexes are separate. Fertilization is external. Depending on species, eggs may be retained within the female until they hatch, or deposited in leathery capsules that hatch either as planktonic larvae or crawling young.

NEOGASTROPODA—MURICOIDEA—MURICIDAE
MUREX SHELLS, ROCK SHELLS

The family Muricidae is a very large and highly variable lineage of predatory mollusks that have evolved the ability to bore holes through the shells of their prey by applying secretions from a specialized gland on the sole of their foot and removing shell particles with their radula.

Most muricid shells have a pronounced sculpture of axial ribs and periodic varices (indicative of episodic growth) with simple to elaborately complex spines or web-like expansions. Spiral sculpture of cords and furrows is also common. Animals living on intertidal rocks tend to have smoother shells and less-pronounced sculpture. Apertures are ovate, with an anterior siphonal canal that can be open or closed, short to very long, and highly sculptured with spines or frilled edges. A thick, corneous operculum

RIGHT | *Drupa rubusidaeus* on shallow reef in Kwajalein Atoll, Marshall Islands.

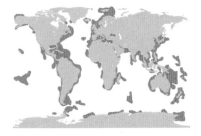

DISTRIBUTION
Benthic marine habitats worldwide, ranging from intertidal to abyssal depths. Some are present in estuaries.

DIVERSITY
Family includes approximately 1,700 living species assigned to 176 genera within 13 subfamilies.

HABITAT
From rocky shores to sand and mud bottoms. Most species are free-living, though some are specialized predators of corals, embedding themselves in a colony for their entire life.

SIZE
Species range in size from ¼–12 in (6–300 mm).

DIET
Carnivores that feed on gastropods, bivalves, barnacles, polychaetes, other invertebrates, and carrion.

ABOVE | Dorsal view of the shell of *Chicoreus palmarosae*, an epifaunal species living on offshore rocky bottoms. Shells from Sri Lanka tend to have purple or pink fronds, while those from the Philippines have shorter brown fronds.

REPRODUCTION
Sexes are separate and fertilization is internal. Eggs are deposited in corneous capsules that vary in shape and have preformed exit plugs that dissolve when the larvae are ready to hatch. Many species hatch as crawling juveniles; others as pelagic larvae that spend time in the plankton. Members of the subfamily Coralliophilinae are protandrous hermaphrodites. Females brood eggs within their mantle cavity.

is present and fills the aperture when the animal is retracted.

Anatomical features of most muricids include a ventral pedal gland that produces secretions to dissolve the shells of their prey, a distinctive radula and smaller accessory salivary glands, and a very well-developed hypobranchial gland that produces a purple secretion. Several Mediterranean species had been harvested since ancient times to produce a prized purple pigment (Tyrian purple), used to dye togas during the Roman Empire.

A few larger species are eaten by humans, but most muricids are regarded as pests that feed on commercially harvested bivalves. Several species have also been used in traditional medicine for wound healing and stomach distress, among other symptoms. Some intertidal species are studied as bioindicators of various environmental pollutants that cause deformities in the snails.

Earliest records attributed to the Muricidae date to the Latest Lower Cretaceous (Albian 100–113 mya).

LEFT | Apertural view of the shell of *Murex pecten*, a species that lives on soft bottoms throughout the Indo-West Pacific. The spines provide protection from predators and prevent the shell from sinking into the mud.

NEOGASTROPODA—TURBINELLOIDEA—TURBINELLIDAE
CHANK SHELLS, VASE SHELLS

BELOW | *Vasum cassiforme*, a species endemic to Brazil, inhabits offshore waters. Species from calm waters tend to have longer spines.

Members of the family Turbinellidae are specialized predators of tube-dwelling worms, including polychaetes and sipunculids. They have evolved an exceptionally long and narrow proboscis that they insert into burrows to extract their prey. The family is currently subdivided into two subfamilies, Turbinellinae and Vasinae, with the closely related Columbariidae, which are similar in diet and several morphological characters, having recently been elevated to the rank of a separate family.

The subfamily Turbinellinae includes two genera, each with large, heavy biconic shells, ovate apertures, and elongated siphonal canals. The genus *Syrinx* contains *Syrinx aruana*, which is the largest living gastropod species (up to 40 in / 1 m in length). It is endemic to Northern Australia and Indonesia. In contrast, the genus *Turbinella* has several species, some restricted to central Indian or western Indian Oceans; others more widely distributed throughout the tropical western Atlantic. All inhabit shallow-water habitats. The species *Turbinella pyrum* is regarded as sacred in India. In Hindu mythology, the god Krishna is often represented holding a sinistral specimen of this

DISTRIBUTION
Tropical and temperate latitudes.

DIVERSITY
Family includes 40 living species assigned to 7 genera within 2 subfamilies.

HABITAT
Sand and rubble bottoms from the intertidal zone to bathyal depths of 656 ft (200 m). Juveniles may burrow, but adults are epibenthic.

SIZE
Species range in size from 1–36 in (25 mm–1 m). Species of the subfamily Vasinae are smaller, some with an adult size of 1–2 in (25–50 mm).

DIET
Predators of tube-dwelling polychaetes, sipunculans, and gaping bivalves.

LEFT | *Turbinella pyrum*, crawling on sand exposed at low tide, Maharashtra, India.

BELOW | Shells of *Syrinx aruana*, the largest species of gastropod, which can reach 40 in (1 m) in length. They inhabit sandy bottoms off Northern Australia and Papua New Guinea.

species. Trumpets made from this species continue to be used in ceremonies and extremely rare sinistral specimens of this normally dextral species are greatly prized.

The subfamily Vasinae is more diverse, and its members are more varied in size, morphology, and ecology. Shells may be thick, heavy and conical, or thin and elongated, with the spire ranging from tall to nearly flat depending on genus. Siphonal canals may be short or span more than half the length of the shell. The genus *Vasum* occurs primarily in shallow waters while *Tudivasum* occur in offshore depths, generally 30–650 ft (10–200 m). Earliest fossils of Turbinellidae are from the Late Cretaceous (Campanian 83.6–72.1 mya).

REPRODUCTION
Sexes are separate. Fertilization is internal. Females deposit eggs in proteinaceous capsules. Capsules of Turbinellinae are large and deposited in strands. Capsules of Vasinae are individually attached to hard substrates. All capsules contain nurse eggs; juveniles hatch as crawling young.

NEOGASTROPODA—OLIVOIDEA—OLIVIDAE
OLIVE SNAILS

Olive Snails are readily recognized by their distinctive elongate, cylindrical shells with a low spire and a very long, narrow aperture with a thickened columellar region with pronounced teeth plaits and a broad siphonal notch at the anterior margin. The exterior of the shell has a glossy surface, lacks a periostracum, and often exhibits complex color patterns. Species of the subfamily Olivinae lack an operculum, while the other subfamilies may have a thin, corneous operculum.

The broad, well-developed foot has a crescent-shaped propodium. The mantle has broad lobes that can envelope the shell, producing the glossy surface. The head is small, with reduced tentacles and eyes. The siphon is long and narrow, and used to detect prey. Olive Snails are capable of very rapid burrowing. The larger species feed on a wide variety of invertebrates as well as on carrion. Prey is captured by the propodium and passed into a pouch formed in the sole of the foot, where it may be

DISTRIBUTION
Shallow waters of tropical seas throughout the world, and also temperate regions, primarily in the southern hemisphere.

DIVERSITY
Family includes approximately 350 living species assigned to 11 genera within 5 subfamilies.

HABITAT
Sandy substrates at intertidal and subtidal depths along the continental shelf.

SIZE
Species range in size from ½–5 in (12–125 mm).

DIET
Carnivores that feed on other snails, infaunal bivalves, invertebrates, and carrion.

OPPOSITE | *Oliva sayana* eating a dead crab (*Aranaeus cribrarius*) on intertidal sand flats off Sanibel Island, Florida.

ABOVE | Multiple views of *Oliva guttata*, a common shallow-water species from the tropical Indo-West Pacific.

REPRODUCTION
Sexes are separate. Fertilization is internal. Females deposit eggs in gelatinous capsules that lay freely on the surface of the sand or are attached to the bottom. Eggs hatch to release free-swimming larvae in three to five days.

partially digested by enzymes secreted by the epithelium of the sole. Some species are capable of autotomizing the posterior portion of their foot as a defense mechanism to escape predators but must regenerate the foot before being able to feed again. Smaller species feed on smaller invertebrates and foraminifera. When inactive, Olive Snails burrow in the sand, but emerge at night and with an incoming tide.

Several species of *Oliva* were recently reported to contain toxins that contribute to paralytic poisoning. Olividae first appear in the fossil record during the Late Cretaceous (Campanian 83.6–72.1 mya).

NEOGASTROPODA—MITROIDEA—MITRIDAE
MITERS

Miter Shells are a distinctive group of neogastropods readily recognized as having a solid, fusiform-to-cylindrical shell with a pointed spire (hence the common name), a narrow, elongate aperture with multiple, prominent columellar folds, a smooth outer lip, and an anterior siphonal notch. Surface sculpture, when present, is spiral, and may consist of spiral grooves, cords, rows of nodules, or pits. Mitridae lack an operculum. Many

BELOW | *Roseomitra rosacea* on a shallow reef off Cebu, the Philippines.

DISTRIBUTION
Occurs primarily in shallow seas globally, in tropical and temperate latitudes, with highest diversity in the tropical Indo-Pacific.

DIVERSITY
Family includes approximately 400 living species assigned to 38 genera within 7 subfamilies.

HABITAT
Most are infaunal, burrowing below the surface of sandy bottoms in shallow waters. Others live on or under corals and rocks from the intertidal zone to depths of over 1,000 ft (300 m).

SIZE
Species range in size from ¼–7 in (5–170 mm).

DIET
Peanut Worms (Sipuncula), which they extract from their burrows.

Indo-Pacific species are brightly colored, often with complex patterns of spiral bands or patches of color, or tent-like patterns, while eastern Pacific species are drab, often with dark shells.

All Mitridae have distinctive anatomical features of their alimentary system that distinguish them from all other neogastropods. These features, which are associated with their highly specific diet that is limited to Peanut Worms (Sipuncula), include an evertible epiproboscis, a complex muscular organ that extends beyond the buccal mass and contains a duct from the salivary glands, and a radula with broad, multicuspid lateral teeth. Mitridae have lost their accessory salivary glands and gland of Leiblein. Their hypobranchial gland produces a purple secretion. Animals have a broad, crawling foot that is often brightly pigmented with spots or stripes.

The earliest fossil record of members of Mitridae dates to the base of the Late Cretaceous (105.3 mya), with an adaptive radiation of most of the modern genera, all with the adaptation to the specialized diet, occurring during the Miocene (20 mya).

REPRODUCTION
Sexes are separate. Fertilization is internal, with females depositing fertilized eggs in vase-shaped proteinaceous capsules attached to the substrate by a stalk. Juveniles of some species hatch as free-swimming veliger larvae; other species undergo direct development within the capsules and hatch as crawling juveniles.

ABOVE | *Isara carbonaria* depositing eggs in capsules on a shallow reef off Australia.

LEFT | Apertural view of the shell of *Mitra mitra*, a common shallow-water species from the tropical Indo-Pacific, showing prominent columellar folds.

NEOGASTROPODA—CONOIDEA—TURRIDAE
TURRID SHELLS

The family Turridae had been considered the largest and most diverse molluscan family, with more than 10,000 species and 600 genera. With origins in the Late Cretaceous (Coniacian 89.8–86.3 mya), it is the oldest among the families within Conoidea and recognized as including a series of intermediate anatomical stages in the development of the hypodermic toxoglossan radula and the numerous toxins produced in the poison gland that are present in Conidae and Terebridae. Based on recent phylogenetic studies, researchers have partitioned the many taxa that had been included in Turridae into more than a dozen separate families, among them a narrower and more circumscribed concept of the family Turridae.

The Turrid Shell is typically elongated, with a tall spire that is conical and comprises half or more of the shell length. The aperture is narrowly ovate, with an axial siphonal canal that may be long or short, and an indentation or slit at the posterior margin of the outer lip below the suture that accommodates

RIGHT | *Lophiotoma picturata* crawling on a reef in a lagoon on Kwajalein Atoll, Marshall Islands.

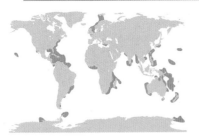

DISTRIBUTION
Global distribution, from equatorial to polar regions, with greatest diversity at tropical latitudes.

DIVERSITY
Family includes approximately 200 living species assigned to 15 genera within 1 subfamily.

HABITAT
Many live on coral reefs and sandy bottoms along the continental shelf; others live at bathyal and abyssal depths. Species within genera may occupy different depths along the same coastline.

SIZE
Species range in size from ½–6 in (12–150 mm).

the exhalant canal. The shell surface has spiral ribs or cords that vary in prominence. Some may have nodules along the shoulder. The operculum is thin, corneous, has an apical nucleus and conforms to the shape of the aperture.

Animals have a long, narrow foot, a head with an elongated rostrum and tentacles with eyes on bulges along their distal outer surfaces. The radula of Turridae has a well-developed membrane with three teeth per row. The venom gland is long with a muscular bulb at its end. Animals are active predators of worms, primarily polychaetes, but also sipunculans and nemertians. Species living in sympatry with other turrids may specialize in different prey species.

ABOVE | Dorsal and lateral views of the shell of *Turris crispa*, the largest species within Turridae, from the tropical Indo-West Pacific. The exhaling current leaves the mantle cavity through the pronounced slit along the shoulder.

RIGHT | Two views of the shell of *Gemmula kieneri*, an offshore species occurring throughout the Indo-West Pacific, where it feeds on terebellid worms.

DIET
Carnivorous, preying primarily on polychaete worms.

REPRODUCTION
Sexes are separate. Fertilization is internal. Eggs are deposited in leathery capsules. Some hatch as crawling juveniles, others as planktonic larvae; some with a long feeding stage before settling to the bottom.

NEOGASTROPODA—CONOIDEA—TEREBRIDAE
AUGER SHELLS

Auger Shells are typically very tall, conical, and slender, with some species having as many as forty whorls. The surface may be smooth, have multiple axial ribs, spiral bands, or with cancellated sculpture. The color may be uniform or variously patterned with bands or spots.

The aperture is small and oval with a broad siphonal notch and the columella may be smooth or have one or two plaits. The outer lip lacks denticles or lirae. The operculum is corneous, with an apical nucleus. Animals have a long foot that is rounded anteriorly and tapers posteriorly. Some intertidal species can use their foot as a water sail to ride receding waves to deeper water. Eyes are present at the tips of short tentacles. The proboscis is thin and long. Some genera have an outgrowth within the rhynchodeum (accessory rhynchodeal lobe), which may function to locate prey.

There are three types of feeding mechanisms within this family. The most prevalent is similar to that of the Conidae, in which animals use a hollow, barbed radular tooth to inject venom to paralyze prey that is then consumed. Other genera have a radula with solid, recurved teeth but lack a poison

LEFT | *Hastula solida* on sandy bottom in about 30 ft (10 m) of water on Kwajalein Atoll, Marshall Islands.

DISTRIBUTION
Widely distributed in tropical and subtropical waters around the world, with highest diversity in the tropical Indo-Pacific.

DIVERSITY
Family includes approximately 400 living species assigned to 21 genera within 3 subfamilies.

HABITAT
Sand dwelling, burrowing in sand to a depth not exceeding the shell length. Some in subtidal coral reefs; others on surf zones of beaches. Majority intertidal to 130 ft (40 m). Others subtidal to 1,150 ft (350 m), mostly in temperate latitudes. They can be abundant in some habitats.

SIZE
Species range in size from ½–10 in (12–250 mm).

FAR LEFT | *Duplicaria dussumierii* from the northwestern Pacific Ocean is among the species that lacks a poison gland and captures its prey with it recurved radular teeth.

CENTER LEFT | *Hastula lanceata*, from shallow tropical waters of the Pacific, is among the species that uses a barbed radular tooth and venom to capture its prey.

LEFT | *Neoterebra armillata*, a shallow-water species from the tropical Eastern Pacific that lacks a radula and poison gland. It captures and swallows small worms whole.

BELOW | *Terebra babylonia* is a species that inhabits subtidal sand patches throughout Central Pacific lagoons.

BELOW | *Terebra triseriata*, an extremely elongated species that may have forty or more whorls, lives on offshore muddy bottoms throughout the Indo-Pacific.

gland, instead capturing and rasping their prey. A third group lacks both radula and poison gland, capturing and swallowing small worms whole.

Terebridae first appear in the fossil record during the Lower Eocene (Ypressian 56–47.8 mya). Diversification of the genera likely took place in the Oligocene and Miocene (33.7–5.3 mya).

DIET
Carnivores, feeding on polychaetes and enteropneusts (acorn worms and hemicordates).

REPRODUCTION
Sexes are separate. Fertilization is internal. Females deposit eggs in stalked capsules that are attached to stones or other objects such as buried shells. Development is direct with crawling juveniles emerging from the capsules.

NEOGASTROPODA—CONOIDEA—CONIDAE

CONE SHELLS

Cone Shells are easily identified by their shape. The spire ranges from tall and conical to nearly flat. The last whorl of the shell is conical, with a sharp shoulder and an aperture that is long, narrow, and has nearly parallel sides, although the apertures of fish-eating species tend to be broader anteriorly. The shell exterior is usually smooth or with fine spiral sculpture. The operculum is corneous, narrow, and spans only a small portion of the aperture length. Many species have elaborate and complex color patterns. The internal walls of previous whorls are often resorbed and become very thin.

Members of the family Conidae are carnivorous marine snails that have evolved a highly specialized feeding mechanism and behavior, injecting their prey with a powerful venom through a hollow, barbed, harpoon-like radular tooth that is held at the tip of an extensible proboscis. The prey is rapidly paralyzed and then swallowed.

The venom, produced by a specialized venom gland, is composed of over a hundred separate neurotoxic compounds, mostly short peptides called conotoxins. Each species within the family produces a unique complex of peptides. Within Conidae, most species feed on worms or snails and their venoms are not especially dangerous to humans. However, fish-eating conids produce venom that has caused more than thirty recorded human fatalities.

The venom of *Conus geographus* is especially toxic and can kill a human in five minutes or less. This species is locally called the "cigarette snail" as the victim is said to only have time to smoke one cigarette before dying. A great deal of research has been devoted to isolating, characterizing, and synthesizing conotoxins, with one drug already approved as a painkiller (reported to be 1,000 times more powerful than morphine).

Notable anatomical novelties include modification of the radula into a series of individual, hollow tubes with

LEFT | *Conus gloriamaris*, the Glory of the Sea Cone, once a great rarity, from offshore reefs in the Philippines.

DISTRIBUTION
Occurs primarily in tropical seas, with greatest diversity in the Indo-Pacific region. Some species extend into temperate latitudes.

DIVERSITY
Family includes approximately 1,300 living species assigned to 9 genera within 1 subfamily.

HABITAT
A variety of habitats including sandy bottoms, coral reefs, and rocky shores, from the intertidal zone to depths of up to 1,600 ft (500 m). Most are nocturnal, burrowing during the day and emerging to feed at night.

SIZE
Species range in size from ¾–9 in (19–226 mm).

one or more barbs at the anterior end, the development of a venom gland with a terminal muscular bulb that contracts to inject the venom, and a long proboscis that emerges from a labial tube (rhynchodeum) that envelopes the paralyzed prey.

The family Conidae is first recorded from Paleocene deposits (Thanetian 59.2–56 mya) and radiated rapidly during the Eocene and Miocene.

ABOVE | *Conus textile* captures its prey by injecting venom through a hollow, harpoon-like radular tooth at the tip of its extended proboscis.

BELOW RIGHT | *Conus striatus*, a common, fish-eating cone that occurs in shallow waters througout the tropical Indo-West Pacific Ocean.

RIGHT | *Conus marmoreus*, a common, shallow-water cone from the tropical Indo-Pacific.

DIET
Carnivorous predators. Some feed on polychaetes, others on mollusks, and some on fish. All hunt and capture living prey.

REPRODUCTION
Sexes are separate. Fertilization is internal. Females deposit eggs in proteinaceous egg capsules attached to hard substrates or anchored within soft bottoms. Most species have a planktotrophic larvae stage.

HETEROBRANCHIA—ARCHITECTONICOIDEA—ARCHITECTONICIDAE
SUNDIAL SHELLS

Architectonicidae are marine snails that have developed specialized adaptations of their alimentary systems for feeding on cnidarians such as corals.

Their distinctive shells may be low-spired or conical and have a wide umbilicus with the anastropically coiled protoconch visible at its terminus. The adult shell is dextrally coiled, but the protoconch is sinistrally coiled, with the direction of coiling reversing along the coiling axis when the pelagic larva settles to the ocean bottom and begins a benthic adult life. The shell sculpture varies among genera and species. It may be smooth in some species, but more usually consists of spiral cords of varying strength that may be beaded and separated by grooves or channels. The operculum is corneous, and ranges from thin to solid, conical and spirally grooved. The inner surface has a small peg.

The animal has a broad foot with lateral extensions along the anterior end. The head has a pair of long tentacles with eyes along their outer bases. The eyes have large lenses. The proboscis may be extended to a considerable length. The radula is composed of a central tooth with multiple cusps that is flanked by a variable number of curved, cusped marginal teeth. The jaws are long and

DISTRIBUTION
Tropical and temperate shores, ranging in depth from subtidal to nearly 3,280 ft (1,000 m).

DIVERSITY
Family includes approximately 120 living species assigned to 12 genera within 1 subfamily.

HABITAT
Tend to be cryptic, burrowing in sand or remaining under rocks, and emerge at night to feed.

SIZE
Species range in size from ½–2½ in (13–64 mm).

DIET
Ectoparasites on colonial cnidarians, primarily corals.

OPPOSITE | Apical and apertural views of the shell of *Heliacus areola*, one of the smallest architectonicids, which ranges from eastern Africa to the Central Pacific.

ABOVE | *Architectonica perspectiva* crawling along sand bottom in 16 ft (5 m) of water off Culasi, Panay Island, the Philippines.

REPRODUCTION
Sexes are separate in some species; others are protandrous or simultaneous hermaphrodites. Sperm is transferred via spermatophores (packets of sperm), and fertilization is internal. Eggs are deposited in gelatinous masses, and larvae may remain in the plankton for several months before settling to a benthic habitat.

narrow, and the buccal cavity and esophagus are lined with cuticle.

Architectonicidae lack ctenidia (gills) and respire through epithelial lamellae formed along the hypobranchial gland. The large pallial kidney and epithelium along the mantle roof also function in gas exchange.

Earliest records of Architectonicidae are from Triassic deposits (Ladinian 237–242 mya).

PYLOPULMONATA—PYRAMIDELLOIDEA—PYRAMIDELLIDAE
PYRAM SHELLS

The Pyramidellidae comprise an extremely diverse family of small to very small marine snails adapted for an ectoparasitic mode of life. Despite their small size, they are often among the most abundant families in many habitats.

Shells are conical or turreted, mostly tall and slender (but some are wider than tall) and may be smooth and glossy or have axial ribs and/or spiral cords. Colors range from white to shades of tan or pink, often with darker brown bands or spots. A distinctive feature of pyramidellids is their heterostrophic protoconch. The larval shell (protoconch) is coiled in the opposite direction (sinistrally coiled) compared to the adult shell, which is dextrally coiled. The protoconch is deflected from the coiling axis at the apex of the adult shell, usually by either 90 degrees or 180 degrees, depending on species. The aperture is oval to ovate. The columella may be smooth or have one or more plaits that may be situated deep within the aperture. The operculum has few whorls, a terminal nucleus, and a ridge along the outer edge of the inner surface.

Animals have a lobed structure (mentum) between the foot and the front of the head. Eyes are

RIGHT | *Styloptygma aciculina* crawling along a shallow reef off Kwajalein Atoll, Marshall Islands.

DISTRIBUTION
Global, however individual species and genera may have more geographically restricted distributions.

DIVERSITY
Family includes approximately 6,000 living species assigned to approximately 350 genera within 4 subfamilies.

HABITAT
Largely conform to habitats of their primary hosts. Most occur in intertidal or subtidal mud bottoms inhabited by various species of worms. Some have been reported from abyssal depths.

SIZE
Species range in size from $1/64$–$1\frac{1}{2}$ in (0.5–35 mm). Most are less than $\frac{1}{2}$ in (12 mm).

at the base of tentacles that are grooved and have a concave surface. Animals lack a radula. The jaw forms a hollow stylet at the end of the proboscis that is used to pierce the skin of its host and suck body fluids.

Pyramidellids have been reported to infest beds of commercially important bivalves, among them oysters, mussels, and giant clams.

Early records of Pyramidellidae appear in fossil beds of the Uppermost Cretaceous (Maastrichtian 72.1–66 mya). Many of the genera diversified during the Tertiary.

ABOVE | *Longchaeus turritus* inhabits shallow reefs in the tropical Indo-Pacific.

RIGHT | *Pyramidella terebelloides* occurs throughout the Islands of the tropical Indo-Pacific. The aperture is reinforced to resist crushing predators.

DIET
Ectoparasites that feed on invertebrates, including mollusks, polychaete, annelid, and other worms, as well as crustaceans, anemones, and their primary hosts.

REPRODUCTION
Pyramidellidae are simultaneous hermaphrodites. Fertilization is internal via transfer of spermatophores. Females deposit eggs in mucous envelopes often attached to their hosts. Eggs are connected by calazae. Most species hatch as pelagic larvae.

ACTEONIMORPHA—ACTEONOIDEA—APLUSTRIDAE
PAPER BUBBLE SHELLS

The family Aplustridae (formerly known as Hydatinidae) encompasses a lineage of epifaunal opisthobranchs with shells that are reduced in relative size and thickness. The animals are voluminous and cannot completely withdraw into their shells.

The shell is very thin, globose to ovate, with a sunken to slightly elevated spire depending on species. The suture between whorls is present and well-defined. The shell surface is smooth with very fine growth striae and a thin periostracum. The last whorl is very large and often conspicuously colored with spiral or axial patterns. The aperture is broadest anteriorly, the outer lip is thin, and the columella is smooth with a thin callus anteriorly.

The animals are large and brightly colored. The foot is large with the lateral edges extending broadly beyond the sides of the shell. An operculum is not present in adults. The cephalic shield has four tentacular processes in front flanking a pair of eyes, and two lobes extending posteriorly over the anterior portion of the shell. The posterior region of the

DISTRIBUTION
Most tropical and subtropical oceans, though not reported from the eastern Pacific. Some species range from the tropical western Atlantic throughout the Indo-Pacific to Hawai'i.

DIVERSITY
Family includes 31 living species assigned to 6 genera within 1 subfamily.

HABITAT
Epifaunal, inhabiting sand, mud, and seagrass bottoms as well as coral reefs at intertidal to shallow subtidal depths generally less than 100 ft (30 m).

SIZE
Species range in size from ½–2½ in (12–65 mm).

OPPOSITE | An atypical color form of the shell of *Aplustrum aplustre* from Reunion Island, Indian Ocean.

ABOVE | *Hydatina physis*, collected in 49 ft (15 m) of water off Queensland, Australia.

DIET
Polychaete worms (Amphinomidae and Cirratulidae), from which they sequester toxins that they incorporate into their own tissues to defend against predators.

REPRODUCTION
Hermaphrodites. Fertilization is internal and mating is reciprocal. Eggs are deposited in twisted ribbons that are anchored in the sand at one end. They develop as veliger larvae.

mantle extends to form an infrapallial lobe that extends over the apex of the shell. A large plicate gill is present and occasionally extends beyond the mantle cavity. The animal everts a large proboscis to capture its prey of polychaete worms. The radula may lack central (rachidian) teeth, but has twenty-six to forty-two lateral teeth per row, each with multiple cusps.

Fossils tentatively identified as Aplustridae have been reported from the Upper Cretaceous (Maastrichtian 72.1–66 mya).

CEPHALASPIDEA—HAMINOEOIDEA—HAMINOEIDAE
GLASSY BUBBLE SHELLS

The Haminoeidae is the most diverse family within the order Cephalaspidea in terms of shell morphology. Several genera and species within this family are only tentatively recognized, with their status and relationships uncertain and in need of further study and potential revision.

The shell is very thin, fragile, and translucent, ranging in shape from broadly ovate to cylindrical, and is widest at mid-length. The aperture is broadest anteriorly and spans the length of the shell, with the outer lip extending beyond the sunken spire and columella. The shell surface is generally smooth, but may have very fine spiral striae or threads, and is covered by a thin periostracum. The calcified portion of the inner whorls of the shell may be resorbed in some species, leaving only the periostracum, which may appear as a twisted cord.

The animal, which can expand to become much larger than the shell, can be completely retracted into the shell. The foot is broad and shorter than the shell, with its sides expanded into wing-like parapodia that can envelope the shell and meet along the anterior region of the dorsum. A broad cephalic shield that is rounded anteriorly is used to furrow through the sediments. The posterior edge has lobes with eye spots, but the animal lacks tentacles. An operculum is present in larvae but not in adults.

LEFT | Dorsal view of the shell of *Haminoea hydatis*, with internal whorls visible through the translucent shell.

DISTRIBUTION
All warm or temperate seas. Species are mostly marine, but some can occur in brackish waters.

DIVERSITY
Family includes 130 living species assigned to 17 genera within 1 subfamily.

HABITAT
Sandy or muddy bottoms, mostly close to shore or in bays and estuaries, and in water sufficiently shallow to support photosynthesis of the algae on which they feed. Some are infaunal during the day and emerge to feed at night.

SIZE
Species range in size from 1/16–1 1/4 in (2–32 mm) in length. Most are about 1 in (25 mm).

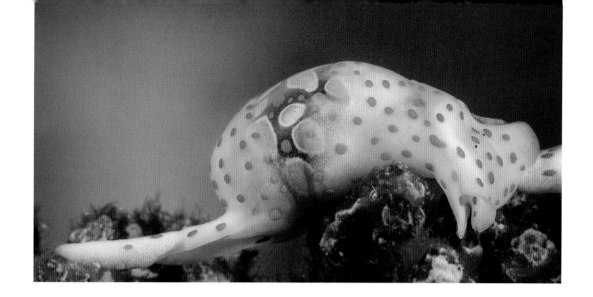

The mantle cavity contains a large gill and both male and female reproductive tracts in the same animal. The buccal mass is large and muscular, with chitinous jaws and gizzard plates, and a radula with multiple hook-shaped teeth per row. Haminoeidae can be common and occur in groups. Earliest Haminoeidae have been reported from mid-Lower Cretaceous deposits.

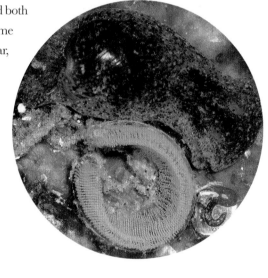

ABOVE | *Haminoea cymbalum* crawling along a reef off Fiji.

ABOVE RIGHT | Animal of *Haminoea antillarum* with egg mass on seagrass flats off Sarasota Bay, Florida.

RIGHT | Dorsal and apertural views of a shell of *Haminoea antillarum* from off Sanibel, Florida.

DIET
Most species graze on green algal turf growing on mud.

REPRODUCTION
Hermaphroditic, with both male and female reproductive tracts in the same animal. Copulation is reciprocal. Eggs are deposited in gelatinous, collar-shaped strands attached to vegetation or sand and containing thousands of eggs.

APLYSIIDA—APLYSIOIDEA—APLYSIIDAE
SEA HARES

BELOW | *Aplysia fasciata* swimming over a seagrass bed in the Mediterranean Sea.

Sea Hares are large animals with fleshy bodies that have a broad foot, a pair of wing-like parapodia that can be used to swim through the water, and a distinct head, with a pair of tentacles and two large elongated and erect rhinophores—chemosensory organs that resemble the ears of rabbit—hence the common name. Sea Hares have internal shells located on the animal's dorsum, beneath the mantle, covering the gill and visceral organs. Shells are relatively small compared to the animal, ear-shaped, and flattened. The shell may be moderately thick, calcified and slightly coiled in some genera, thin and translucent in others, and lost entirely in the post-larval stage of several genera.

Aplysiidae inhabit shallow waters in tropical and temperate regions, feeding on algae and seagrasses, and growing rapidly. Despite a short life span of about one year, some of the larger species may grow to more than 2 ft (0.6 m)

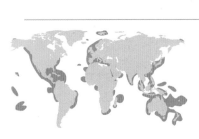

DISTRIBUTION
Cosmopolitan throughout tropical and temperate latitudes, primarily in shallow areas along coasts and bays.

DIVERSITY
Family includes over 80 living species assigned to 10 genera within 1 subfamily.

HABITAT
Shallow water, on both rocky and sandy substrates such as seagrass beds. Animals tend to be nocturnal.

SIZE
Species range in size from 1–29 in (25–750 mm).

DIET
Herbivorous, feeding on a variety of algae and seagrasses.

ABOVE LEFT | Dorsal view of the internal shell of *Aplysia dactylomela* from the Bahamas.

ABOVE RIGHT | Dorsal view of the internal shell of *Aplysia brasiliana* from the Matanzas River Inlet, Florida.

in length and weigh over 30 lbs (13 kg). They derive their color from pigments in the algae they consume, and concentrate toxins produced by the algae. When disturbed, they release a purple ink as a means of defense.

Aplysia has only about 20,000 neurons, and has been used as a model organism for numerous studies by neurobiologists. A study on the molecular basis of learning and memory using *Aplysia californica* earned a Nobel Prize.

Early fossil records of Aplysiidae are from the Lower Miocene (Burdigalian 20.4–16.0 mya).

RIGHT | *Bursatella leachii* crawling along the bottom in shallow water off the south shore of Hong Kong, China.

REPRODUCTION
Simultaneous hermaphrodites with functional male and female organs. Fertilization is internal. Eggs are shed in long strings and hatch as planktonic larvae. Life span of about one year and reproduce only once in their lifetime.

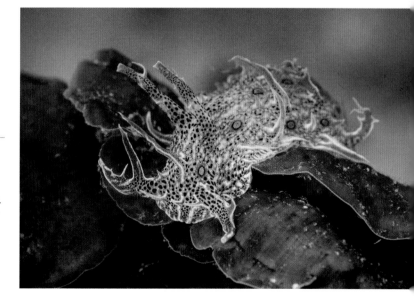

PTEROPODA—CAVOLINIOIDEA—CAVOLINIIDAE
SEA BUTTERFLIES, PTEROPODS

Sea Butterflies spend their entire lives floating and swimming through the upper layers of ocean waters, usually in the top 330 ft (100 m). They swim by flapping their two broad wings (parapodia) that are formed from the anterior portion of the foot that protrudes from the aperture of the shell.

The shell is very thin, translucent, and bilaterally symmetrical, with distinctly different dorsal and ventral surfaces. It may be globular, or triangular, tapering posteriorly. The aperture is laterally elongated, beneath the anterior dorsal margin of the shell. There is no operculum.

The animal can completely withdraw into the shell. The head is small, with the mouth situated between the wings and surrounded by three ciliated lobes formed by parts of the foot. A pair of dorsal tentacles emerge from the neck region, the right

DISTRIBUTION
Global, carried by currents through all the oceans of the world.

DIVERSITY
Family includes 38 living species assigned to 4 genera within 2 subfamilies.

HABITAT
Holoplanktonic, mostly at depths of 330–3,280 ft (100–1,000 m), in temperate and warm seas.

SIZE
Species range in size from ¼–1¼ in (6–30 mm).

DIET
Feed primarily on diatoms, foraminiferans, and dinoflagellates captured in their mucus webs.

larger than the left. They produce a large mucus web that traps planktonic organisms such as diatoms, foraminiferans, radiolarians, and dinoflagellates, as well as small particles that are then consumed. A small radula is present. The mantle is large and surrounds the entire shell, and is divided into multiple lobes that increase the surface area and reduce sinking. Gills are present in some but not all genera.

Sea Butterfly shells, as well as those of other pteropods, are very thin and are susceptible to ocean acidification caused by increasing levels of dissolved carbon dioxide. The shells of dead pteropods, including cavolinids, sink to the ocean bottom and form a characteristic pteropod ooze. The earliest fossils of Cavoliniidae are from the Late Oligocene (Chattian 23.0–27.8 mya) deposits.

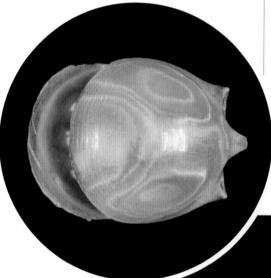

LEFT | Ventral view of the shell of *Cavolinia globulosa* with the aperture on the left.

BELOW | Dorsal and ventral views of the shell of *Diacavolinia longirostris*.

OPPOSITE | *Diacria trispinosa* with egg string attached, swimming over the Mid-Atlantic Ridge, North Atlantic Ocean.

REPRODUCTION
Protandric hermaphrodites, developing from males to hermaphrodites to females. Fertilization is internal. Eggs are deposited in planktonic strands that hatch into veliger larvae.

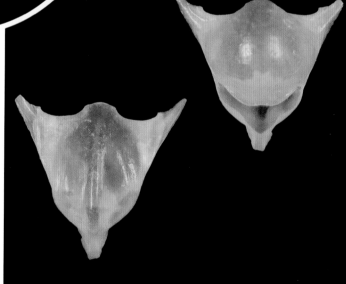

SIPHONARIIDA—SIPHONARIOIDEA—SIPHONARIIDAE
FALSE LIMPETS

BELOW | *Siphonaria gigas* exposed on a rock at low tide along the Pacific coast of Costa Rica.

Siphonariidae are called False Limpets because they are one of over a dozen families that have independently developed limpet-like shells and are not related to true limpets. They cling to intertidal rocks throughout tropical and temperate oceans and have evolved the ability to breathe air when the tide recedes and breathe water when submerged. The roof of their mantle cavity has become vascularized and serves as a lung. A secondary gill within the mantle cavity enables respiration when under water. Air and water enter the mantle cavity through the pneumostome, a respiratory opening along the right side of the body.

Shells are conical and limpet-like or cap-shaped, oval in general outline, but asymmetrical due to a lobe along the right side that indicates the location of the pneumostome. Sculpture consists of radial ribs that are lighter in color than the interstices. The inner

DISTRIBUTION
Tropical, subtropical, and temperate oceans of the world.

DIVERSITY
Family includes 109 living species assigned to 11 genera within 1 subfamily.

HABITAT
Most species live attached to hard substrates such as rocks and seawalls within the range of the intertidal zone. A few species occur above the high-tide line; others below the low-tide line.

SIZE
Species range in size from ¼–2 in (6–50 mm).

ABOVE | Dorsal and ventral views of the shell of *Siphonaria sirius*, a western Pacific species. The gap in the dark-brown C-shaped band on the inner surface of the shell indicates the position of the pneumostome.

DIET
Algae and lichens scraped from the rocks with their radula. Species that live above the high-tide line feed on algae supported by saltwater spray of breaking waves.

REPRODUCTION
Hermaphroditic. Fertilization is by transfer of spermatophores. Eggs are deposited in a gelatinous egg mass that is either attached to rocks or released into the water. Planktotrophic larvae hatch after several days.

surface of the shell is glossy with a slight furrow on the right side marking the siphonal groove that divides the horseshoe-shaped pedal muscle scar. There is no operculum. The foot is large, muscular, and rounded. The head is wide and lacks tentacles.

Many False Limpet species have homing behavior and return to their home scar, to which their shell edges conform precisely, when not feeding, providing protection from predators and waves. False Limpets living in environments exposed to strong wave action tend to have lower, flatter shells than those living in protected areas. Life spans of up to six years have been recorded for some species. Species of some genera live below the low-tide line.

The first occurrence of False Limpets has been reported from the Jurassic (Oxfordian 161.5–154.8 mya).

OPPOSITE | *Ascarosepion latimanus* is a common large cuttlefish along reefs in tropical Indo-Pacific seas, at depths to 100 ft (33 m).

CEPHALOPODA

Cephalopods are the most highly evolved class within Mollusca, and include the most mobile, intelligent, and largest members of the phylum, with the colossal squid reaching 50 ft (15 m) in length and weighing more than 1,000 lbs (450 kg). Cephalopods are bilaterally symmetrical and are elongated along the dorsoventral body axis. They have eight or more arms derived from the molluscan foot that surround the mouth, as well as a radula and a chitinous beak that resembles the beak of a parrot. The mantle is modified and forms a funnel through which water can be forced for propulsion. The skin may contain numerous chromatophores, which allow the animal to change colors and form patterns on its body. Cephalopods have a highly developed nervous system with a centralized brain, the largest of all invertebrates, and complex eyes. Some cephalopods are considered to be intelligent, and capable of learning and remembering.

Sexes are separate. Mating may involve courtship with rapid changes in body color. Males transfer a spermatophore into the mantle cavity of the female using a modified arm. Large eggs with yolk are deposited and juveniles hatch directly without a swimming larval stage. Most cephalopods die after spawning.

Cephalopods are thought to have evolved from monoplacophoran-like ancestors with conical shells. The earliest cephalopods, which appeared during the Upper Cambrian (>500 mya), had external shells with multiple chambers interconnected by a tubular siphuncle that was filled with gas and allowed the animal to control its buoyancy. Several lineages, some with linear shells exceeding 30 ft (9 m) in length, became extinct by the Triassic (200 mya). The majority of cephalopods living today have lost their external shell. Spirula, cuttlefish, and squid all have internal shells, reduced to varying degrees, while octopods lack a shell.

Cephalopods are exclusively marine, inhabiting all oceans of the world at all depths. Some are benthic, ranging from shallow waters to abyssal depths; others are pelagic, swimming or drifting far from shore, surface, or bottom. Pelagic species feed primarily on fish, crustaceans, and other cephalopods, while benthic species feed on a variety of fish and invertebrates.

Living cephalopods range in length from 1 in (25 mm) to 50 ft (15 m) in length. Although more than 15,000 species of fossil cephalopods (nearly all with external shells) have been described, only about 150 genera and 900 species (nearly all without external shells) are living today. Living cephalopods are divided among two subclasses: Nautiloidea, which includes one family, two genera, and seven species, has external shells, while Coleoidea contains all the remaining genera and species. Coleoidea is subdivided into Decapodiformes, the superorder containing squid, which have internal shells, and Octopodiformes, the superorder containing octopuses, which lack shells.

SHELL POSITIONS IN TYPICAL CEPHALOPODS

NAUTILIDAE

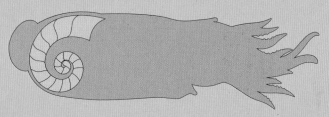

SPIRULIDAE

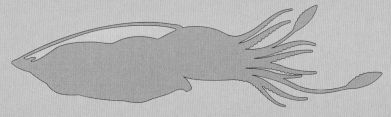

SEPIIDAE

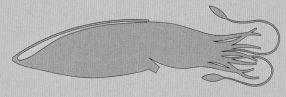

LOLIGINIDAE

NAUTILOIDEA — NATILIDA — NAUTILIDAE
NAUTILUS

Species of Nautilidae are the only living cephalopods with external shells. The shells are bilaterally symmetrical, planispirally coiled and chambered, with the body of the animal occupying the last open chamber. The other chambers contain gas and liquid, and are interconnected by a siphuncle, which the animal uses to control buoyancy. Shells have fine growth striae and are cream colored with reddish-brown radial markings that fade near the aperture. The region of the shell adjacent to the aperture is dark brown or black and is where the hood of the animal touches the shell. The inner surfaces of the shell are nacreous.

The heads of living species of Nautilidae are covered by a fleshy hood that can serve as a shield. The mouth is surrounded by up to ninety-four tentacles that lack suckers, but each has a terminal filament (cirrus) that can be retracted into a basal

RIGHT | A shell of *Nautilus pompilius*, sectioned to expose the individual chambers and the siphuncle connecting them.

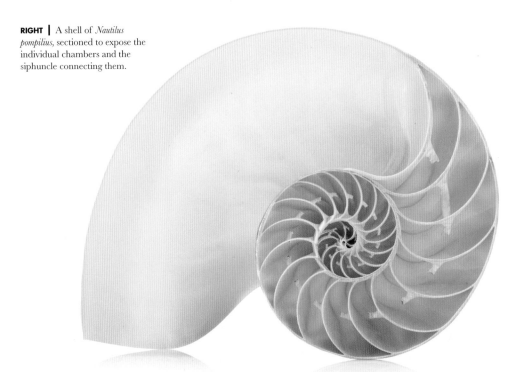

DISTRIBUTION
Tropical Indo-West Pacific Ocean.

DIVERSITY
Family includes 7 living species assigned to 2 genera within 1 subfamily.

HABITAT
Water column where coral reefs drop off into deeper water.

SIZE
Species range in size from 6–10½ in (150–267 mm).

DIET
Feed primarily on shrimp, crabs, and fish.

REPRODUCTION
Sexes are separate. Internal fertilization by transfer of a spermatophore. Female deposits several eggs and attaches them to a hard substrate. Eggs hatch in nine to twelve months. Unlike most other cephalopods, females spawn repeatedly over a period of years.

LEFT | *Nautilus pompilius* swimming in the Coral Sea, off Queensland, Australia.

sheath. The eyes are lateral and of the pinhole type, without a lens. The mantle cavity contains two pairs of gills, lacks an ink sac, and forms a bilobed funnel that can be rolled into a narrow tube. Nautilids may live to twenty years and take five to ten years to reach reproductive maturity. Animals live in deep water (about 1,000 ft / 305 m) adjacent to coral reefs, migrating into shallower waters (300 ft / 90 m) to feed on shrimp, crabs, and fish at night and returning to the depths when the sun rises.

Nautiloidea, the oldest lineage of cephalopods, first appeared in the late Cambrian. All had external shells with multiple chambers. Some were linear (orthoconic) but by the Ordovician, several lineages developed planispirally coiled shells with an inner nacreous layer. All but the family Nautilidae, which is represented in the living fauna by two genera and seven species, became extinct by the end of the Mesozoic.

SEPIIDA—SEPIOIDEA—SEPIIDAE
CUTTLEFISH

The order Sepiida is divided into two suborders, Sepiina and Sepiolina. The former has Sepioidea as the only living superfamily, containing the family Sepiidae.

Species of the diverse family Sepiidae have an oval, dorsoventrally flattened body. The mantle may be slender and elongate to nearly circular, depending on species. All have a calcareous internal shell (sepion) beneath the skin and dorsal to the mantle. Sepiids have eight arms with two to four rows of suckers and two long tentacles that can be retracted into pockets along the ventral sides of the head. Tentacles may have four or more rows of suckers on their broadened distal ends (clubs). Long narrow fins are present along the sides of the mantle and have free lobes at the posterior end. The head is prominent with large eyes that contain a lens and are covered by a transparent membrane. The mouth contains a radula and a beak. The skin contains numerous chromatophores, allowing the animal to rapidly change color and pattern.

The internal shell (sepion) is large, thick, calcareous, chalky white, lanceolate in shape, and nearly flat, with broad lateral margins and a posterior spine. The dorsal surface is expanded; the ventral surface has a wide siphuncle. The shell floats and can be carried by currents before washing ashore.

Like *Nautilus* and *Spirula*, some species of Sepiidae descend to deeper regions during the day and rise to shallower waters to feed at night. The life span of cuttlefish is eighteen to twenty-four months.

LEFT | Dorsal and ventral views of the internal shell of *Acanthosepion pharaonis*, an Indo-West Pacific species.

DISTRIBUTION
Continental shelf and upper slope of temperate and tropical regions, except along the coasts of the Americas.

DIVERSITY
Family includes approximately 120 living species assigned to 13 genera within 1 subfamily.

HABITAT
Animals are benthic to benthopelagic, with most species occurring in shallow coastal waters along the continental shelf. Some species occur at bathyal depths of up to 1,640 ft (500 m).

SIZE
Species range in size from ½–4 in (10–100 mm). Some are nearly 40 in (1 m) including tentacles, with shell and mantle length approaching 20 in (500 mm).

ABOVE | *Acanthosepion pharaonis* in the Andaman Sea off Thailand. The smaller specimen is a male guarding the larger, egg-laying female.

DIET
Crustaceans, mollusks, and fish, hunted primarily at night and in shallower waters.

REPRODUCTION
Sexes are separate. Fertilization is internal, with males introducing spermatophores into the mantle cavity of the female. Soon after mating, 150 to 500 eggs are deposited on hard surfaces, usually in clusters.

MYOPSIDA—LOLIGINIDAE
PENCIL SQUID

Myopsida is one of six orders currently included in Coleoidean superorder Decapodiformes, which contains more than half of all living cephalopod species. All have animals with a pelagic lifestyle, streamlined body with distinct head and fins, eight arms, and two tentacles, and they lack an external shell. Locomotion is primarily by jet propulsion, with the mantle producing a jet of water directed through the funnel that propels the animal. Studies using morphological as well as molecular data have yet to robustly resolve the phylogenetic relationships among these orders.

Myopsida are considered to be close to Sepiida, since both have a cornea in their eyes, suckers with circular muscles, similar beak morphology, tentacles that can be retracted into pockets, and benthic eggs. However, unlike Sepiida, the shell of Myopsida is decalcified and reduced to a long, thin, flexible, chitinous gladius that is situated dorsally within the mantle and lacks chambers or a siphuncle. While sepiids rely on their shells to control buoyancy, Myopsida and other coleoid families

BELOW | Gladii of *Loligo forbesii*.

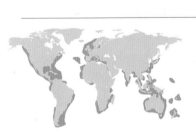

DISTRIBUTION
Diverse and abundant in waters above the continental shelves and bathyal zones along the temperate and tropical coasts of the world's oceans.

DIVERSITY
Family includes 62 living species assigned to 10 genera within 1 subfamily.

HABITAT
Most species are pelagic and occur in large schools, migrating vertically from deeper waters (984 ft / 300 m) during the day to shallower waters (164 ft / 50 m) at night.

SIZE
Species range in size from 1 in (26 mm) to 3 ft (0.9 m).

DIET
Active hunters that feed primarily on fish, crustaceans, and other cephalopods.

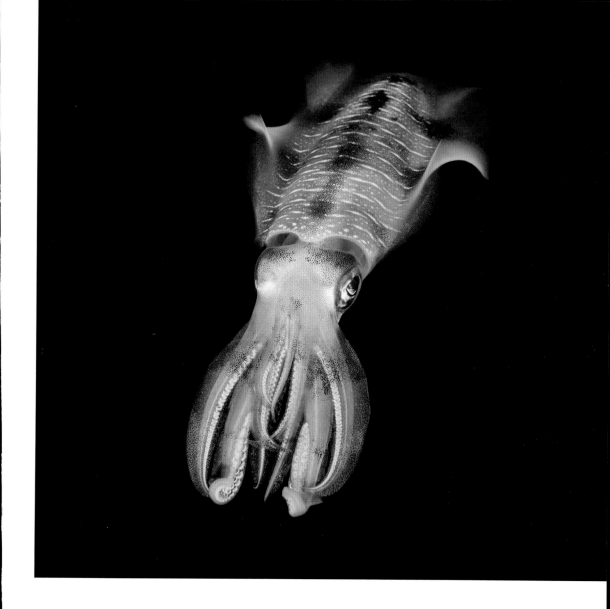

ABOVE | *Sepioteuthis lessoniana* hovering in open water above a coral reef at night off West Papua, Indonesia.

REPRODUCTION
Sexes are separate. The male transfers a spermatophore with a hectocotylized arm into the mantle cavity of the female. Large, yolky, encapsulated eggs are deposited onto the bottom or hard substrates in clusters. Loliginidae have only one breeding season and both males and females die after spawning (semelparous reproduction).

without a calcified shell regulate their buoyancy by storing ammonium chloride in their body tissues.

The order Myopsida includes the family Loliginidae, with ten living genera and sixty-two species, as well as the family Australiteuthidae, with a single, monotypic genus. Several species of loliginids are fished commercially. Courtship behavior, including displays of changes in color and body shape, occurs in many species.

SPIRULIDA—SPIRULOIDEA—SPIRULIDAE
RAM'S HORN SQUID

The order Spirulida contains a single family, Spirulidae, with only a single genus and species, *Spirula spirula*. It is the closest living relative of the extinct orders Belemnitida and Aulacoderida, and this lineage is considered to be closely related to Sepiida (cuttlefish).

BELOW | The internal shell of *Spirula spirula*.

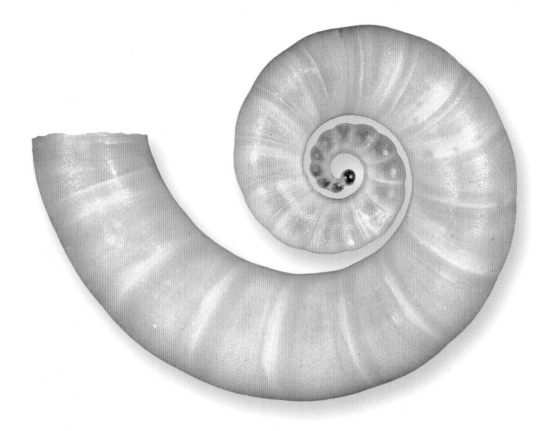

DISTRIBUTION
Deep waters (1,640–3,280 ft / 500–1,000 m) in the tropical Atlantic, Indian, and western Pacific Oceans near the edges of the continental shelf and oceanic islands.

DIVERSITY
Family includes 1 living species, *Spirula spirula*.

HABITAT
They remain in deep water during the day and migrate to shallower water (650–980 ft / 200–300 m) at night to feed.

SIZE
Adults have a mantle length of about 2 in (50 mm). The shell reaches 1 in (25 mm) in diameter.

DIET
Small fish, crustaceans, and other invertebrates.

Like most Coleoidea, *S. spirula* has an internal shell, an ink sac, eight arms, and two tentacles, all with suckers, but is the only member of Coleoidea with a coiled shell. The shell is calcified, planisirally coiled, and gyroconic (adjacent whorls do not touch). It may contain thirty or more chambers that are connected by a siphuncle, a tissue-lined tube through which the animal controls buoyancy. The shell is situated in the posterior portion of the animal, which has a tubular body with a mantle length of about 2 in (50 mm). There are two fins at the posterior portion of the body, and a large photophore (light-emitting organ) between the fins.

The animal swims in a vertical orientation with the head and tentacles directed upward and the fins and photophore directed toward the bottom. The photophore can produce a yellowish-green light for several hours at a time. *Spirula spirula* lack a radula but have a beak, and toxic saliva. Female *S. spirula* are larger than males. The shell is buoyant, will float to the surface after the animal dies, and can be carried for considerable distances by currents before washing ashore.

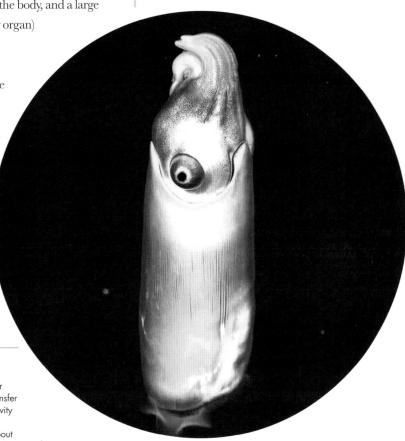

RIGHT | *Spirula spirula*—living animal in 2,790 ft (850 m) of water off the northern Great Barrier Reef, Australia.

REPRODUCTION

Sexes are separate. Females are larger than males, which have their two ventral arms specialized to transfer spermatophores into the mantle cavity of the female. Eggs are laid on the ocean bottom and juveniles are about 1/16 in (2 mm) in size when they hatch. *S. spirula* lives for 1–1.5 years.

OCTOPODA—ARGONAUTOIDEA—ARGONAUTIDAE

PAPER NAUTILUS

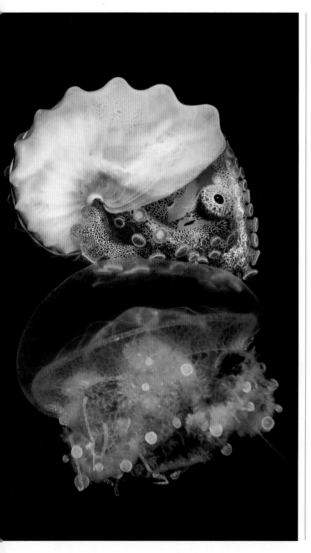

The family Argonautidae contains a single genus *Argonauta* with four living species, each of which produce a large, distinctive, coiled, calcareous "shell" that is not homologous to the shell of other mollusks. Rather than being produced by the mantle, it is secreted by glandular membranes on the dorsal arms of females. The animal is not attached to this "shell," which provides protection for the animal and serves as a flotation device as well as a hatching chamber for multiple batches of eggs. These structures are spirally coiled, laterally compressed, composed of calcite, and consist of a single chamber. Most are whitish, thin, and translucent with two rows of brownish tubercles along the keel that forms its outer edge. The lateral sides have radial ribs that lead to the tubercles. Females may cling to objects floating on the surface, including other *Argonauta*, and chains composed of as many as twenty to thirty individuals have been reported.

All species of *Argonauta* exhibit extreme sexual dimorphism. Females are large and secrete these "shells," while males are less than one-twentieth of the size of females and lack a shell.

LEFT | *Argonauta hians* riding on a jellyfish off Batangas, Luzon, the Philippines.

DISTRIBUTION
Tropical and subtropical seas. Some species have a cosmopolitan distribution; others are limited to the southern hemisphere, or the eastern Pacific Ocean.

DIVERSITY
Family includes 4 living species within 1 genus.

HABITAT
Epipelagic, mostly in the upper 330 ft (100 m) of the ocean.

SIZE
Females produce "shells" that range in size from 1½–12 in (37–300 mm) depending on species. Males lack shells and are less than 1 in (25 mm) in length.

DIET
Feed during the day, primarily on pelagic mollusks such as heteropods and pteropods, as well as crustaceans and jellyfish.

Argonauta species have a radula and an ink sac. The skin of *Argonauta* has chromatophores that can produce rapid changes in color. The eight arms have two rows of small suckers. The two dorsal arms of females have very broad glandular membranes that secrete and hold the delicate shell. The third left arm of males is hectocotylized and detachable. Females may have hectocotylus arms from multiple males within their mantle cavity. The fossil record of *Argonauta* extends to the Oligocene (33–23 mya).

ABOVE | The egg case of *Argonauta cornuta* from off the coast of Baja California, Mexico.

BELOW | A female *Argonauta hians* with dorsal arms enveloping the egg case, in 70 ft (21 m) of water off Anilao, the Philippines.

REPRODUCTION
Sexes are separate. Females deposit multiple cohorts of eggs within the shells, with up to five batches at different stages of development reported. Females with a mantle length of less than 2 in (50 mm) were reported to have 40,000 eggs in the hatching chamber. Hatching usually occurs at night.

APPENDICES

GLOSSARY

aesthetes Sense organs in chitons that penetrate the tegmentum (shell) through numerous pores.

anastrophic When the protoconch is coiled around the same axis as the teleoconch, but with the direction of coiling along the axis reversing at metamorphosis, resulting in the nucleus directed toward the base of the adult shell.

anoxic Without oxygen.

aperture The opening through which the animal extends from the shell, and the most recent part of the shell to be formed.

apex The first portion of the shell to be formed, situated at the tip of the spire.

apophysis A projecting structure to which muscle is attached.

articulamentum The dense inner layer of the polyplacophoran shell valves. Extends beyond the tegmentum and forms the sutural lamellae and insertion plates embedded in the girdle.

auricle An ear-like extension of the hinge line in some bivalves, such as Pectinidae and Pteriidae.

autotomy Self-amputation. The ability of an animal to shed a part of its body, usually the posterior part of the foot in gastropods, to elude a predator. Some animals have the ability to regenerate the lost body part.

axial Parallel to the shell axis.

benthic Living on the bottom in an aquatic environment.

biconic Shaped like two oppositely oriented cones, joined at their bases.

bioturbation The mixing or disturbance of sedimentary deposits by activities of living animals.

bipectinate Refers to gills with two rows of leaflets.

body whorl The last whorl (360 degrees) of a gastropod shell to be deposited.

broadcast spawning The release of both eggs and sperm into the water column. Gametes come into contact and fertilization occurs in the water column.

byssus/byssal Proteinaceous fibers secreted by a gland in the foot of a bivalve used to attach the animal to a hard substrate.

calcarella A type of protoconch or larval shell.

callus A thickened layer of shell along the parietal or columellar region of the shell.

cancellate Having surface sculpture of intersecting axial ribs and spiral cords.

captacula Extensible thread-like tentacles that emerge from the sides of the snout of Scaphopoda. Used to probe the sediment for food particles and retract to bring them to the mouth.

carina A sharp, keel-like ridge.

chalazae String-like links joining egg capsules of some gastropods.

chemoautotroph An organism that derives energy from chemical oxidation of inorganic compounds rather than from sunlight.

chemosynthesis/chemosynthetic A process by which food is produced by bacteria or other living things using energy derived from chemicals rather than sunlight.

chitinous Composed of chitin, a semi-transparent, horny substance found in the shells of some mollusks.

columella The central pillar of the shell, formed by the inner lip of the aperture around the axis of coiling.

commarginal Parallel to the shell margin.

conchiolin The proteinaceous component of a shell.

congener An organism belonging to the same genus as another.

corneous Made of a horn-like substance such as conchiolin or scleroprotein; not calcareous.

coronate Crown-like; having spines, beads, or tubercles along the shoulder of a gastropod shell.

costa (adj. costate) An element of axial sculpture, which may be rounded and rib-like, or flange-like.

ctenidia Respiratory organs or gills present in the

mantle cavities of bivalves, many gastropods, cephalopods, monoplacophorans, and aplacophorans, and in the pallial groove of polyplacophorans, but absent in scaphopods.

ctenolium A comblike row of teeth along the ventral edge of the byssal notch in Pectinidae.

denticle A single, toothlike projection. A shell with such projections is termed dentate or denticulate.

deposit feeder A mollusk that feeds on decomposing organic matter deposited on the floor of its habitat, either on the floor surface or by tunneling deep into it.

dextral Coiled in a clockwise direction when viewed from the apex of a shell; relating to shells with the aperture to the right of the coiling axis.

diatom A type of algae common in the phytoplankton, that is, encased in a cell wall made of silica.

digitation A finger-like projection.

dioecious Individuals are either male or female throughout life.

dissoconch That portion of the shell of a bivalve produced after the larva undergoes metamorphosis; the adult shell.

distal Situated farthest from the base or point of attachment; distant.

ecdysis The process of shedding the outer cuticle (in arthropods).

echinosira A type of larva with a secondary shell that assists with flotation.

epibenthic Living on the surface of sediments in an aquatic environment.

epifaunal Living on the surface of a substrate (for example, rocks, pilings, or other animals) in an aquatic environment.

epipodium/epipodial A muscular expansion of the lateral parts of the foot.

eulamellibranchiate Gills that are symmetrical, W-shaped, connected by filaments, occurring in some bivalves.

fasciole A raised spiral band formed by successive growth lines along the edges of a canal.

fillibranch Gills that are symmetrical, W-shaped, loosely connected, occurring in some bivalves.

filter feed To feed on particles or microorganisms filtered from the water.

flame/flammule An irregular but repeated color-marking on the exterior surface.

fold A spirally wound ridge on the columella of a gastropod.

foliated Having thin, leaflike layers or plates.

foramen The opening at the apex of the bilaterally symmetrical, conical shells of many species, through which water with waste exits the mantle cavity, and through which gametes are released.

Foraminifera A phylum of single-celled microorganisms with a shell formed of calcium carbonate.

fossa A long, shallow, narrow depression in the surface of a shell.

frondose Like a frond or leaf, divided into many sections.

funicle/funiculum A ridge that spirals into the umbilicus of a gastropod shell.

fusiform Spindle-shaped; swollen in the middle, tapering toward the ends.

girdle A band of leathery or muscular tissue holding the valves of a chiton in place.

gladius Internal vestige of a shell present in squid. Composed of chitin.

gonopore A pore or opening through which gametes are extruded.

gorgonian A type of soft coral.

granulose Having a surface covered with granules.

growth line A line on the shell surface that marked the position of the shell margin during an earlier stage of growth; delineate increments of shell growth.

gyroconic Having a coiled conical shell in which the whorls do not touch (see Spirula spirula page 220).

hectocotylized arm/hectocotylus The tip of a specialized arm in coleoid cephalopods used to transfer spermatophores into the female.

heterostrophic Coiling in different directions, referring to protoconch and adult shell (teleoconch) of gastropods.

holdfast A structure that anchors a sessile animal or plant to the substrate. Usually applies to bivalves.

hypobranchial gland A gland on the roof of the mantle cavity that produces secretions that bind sediment for removal from the mantle cavity.

hypostracum Innermost shell layer of polyplacophorans.

imbricate With overlapping scales.

inductura A layer of shell material along the inner lip (parietal region and or columella) of a gastropod shell.

interstice (pl. interstices) The space between two neighboring features such as teeth or ribs.

intertidal The region exposed at low tide and submerged at high tide.

involute A type of shell-coiling in which the last whorl completely envelopes the previous whorls.

keel A sharp, raised, blade-like spiral sculptural element. Usually present along the shell periphery or shoulder.

labrum The outer lip of a coiled shell.

lamella (adj. lamellose) A thin, plate-like structure that often occurs in multiples.

lanceolate Leaf-shaped; narrow and tapering to a point.

lecithotrophic Larvae that do not feed, relying on yolk for nutrition before metamorphosing into juveniles.

lenticular Shaped like a lens, convex on both sides.

lira (pl. lirae) A narrow, linear ridge along the shell, or within its outer lip.

lunule A heart-shaped, depressed region situated in front of the umbones in many bivalves.

maculation A pattern of spots.

mammillate With dome-shaped or nipple-like projections.

mantle The outer portion of a mollusk's body that secretes the shell.

mesoplax An elongated accessory plate formed near the umbo in some pholadid bivalves.

microcarnivore A carnivore that feeds on very small animals/larvae.

monocuspid Having a single cusp or point. Usually refers to radular teeth.

nacre/nacreous An iridescent inner shell layer composed of thin layers of aragonite. Also called mother-of-pearl.

nephridium/nephridia An organ that performs the functions like a vertebrate kidney, removing metabolic wastes from an animal's body.

nepionic Post-embryonic larval stage.

nodulose Having small knobs or nodules.

nudibranch One of a group of gastropods that lack a shell as adults.

ocelli Multiple, light-sensing eyes, as in the mantle margin of a scallop.

operculum A circular or elongated structure produced by some gastropods to block or seal the aperture of the shell when the animal withdraws. May be corneous or calcareous.

opisthobranch A group of gastropods that have their gill posterior to the heart.

oral shield A cuticularized structure ventral to the mouth of Caudofoveata.

osphradium (pl. osphradia) Chemosensory organ adjacent to the gill.

ostracods A class of small crustaceans with a shell composed of two valves.

pallial Pertaining to the mantle.

pallial groove Shallow mantle cavity of polyplacophorans.

pallial line A narrow line or muscle scar that marks the line of attachment of the mantle to the shell in some bivalves.

pallial sinus An indentation in the pallial line marking the attachment of the muscles that retract the siphon in some bivalves.

palp A broad, flattened appendage near the mouth of bivalves.

parapodia Lateral extensions of the foot.

parietal The posterior region of the inner lip of a gastropod, between the columella and the suture.

paucispiral Having few whorls around a central point.

pericalymma A larval type with a ciliated epithelial covering.

periostracum A thin organic layer of conchiolin, the sometimes fibrous outer coating of the shell during the life of many mollusks.

plait A raised fold of shell that occurs on the columella.

planispiral Coiled in a single plane.

plica/plicae A raised fold or ridge.

pneumostome A respiratory opening along the right mantle edge that can be closed.

porcelaneous Having a smooth white surface like fine porcelain.

prodisoconch Larval shell of bivalves. That part of the shell formed before the larva undergoes metamorphosis. Situated at the tip of the umbo.

propodium Anterior portion of a gastropod foot.

protandrous/protandry/protandric A type of hermaphroditism in which an individual begins life as a male and changes into a female.

protobranch A primitive leaf-like gill form in bivalves.

protoconch Larval shell of a gastropod; that part of the shell formed before the larva undergoes metamorphosis. Situated at the tip of the apex.

protoplax An accessory plate covering the anterior dorsal margin of the shells of pholadid bivalves.

psuedolamellibranch Gills that are symmetrical, W-shaped, and joined by filaments, found in some bivalves.

ptenoglossate A type of radula with numerous, similar fang-like teeth in each row.

punctate With small, needle-like depressions.

pyriform Pear-shaped.

radula A flexible ribbon supporting multiple rows of chitinous teeth. A feeding structure unique to mollusks but absent in bivalves.

ramp A shelf-like ledge at the top of a whorl immediately below the suture.

ray A linear mark of surface color, radiating like a ray of light from a central point such as the apex.

resilifer A recess in the bivalve shell where the ligament is attached.

resilium Internal ligament of bivalves.

reticulated Having a netlike pattern.

rhipidoglossan/rhipidoglossate A type of radula in gastropods that includes one central tooth, several lateral teeth and multiple marginal teeth in each row.

rostrum A tapering structure resembling a bird's beak.

rugose Having a surface with wrinkles and ridges.

sagittal Relating to the plane of bilateral symmetry, or any plane parallel to it.

scalariform Resembling a ladder.

seep An area of the ocean floor where cold gases (cold seep) or hot springs (hydrothermal vent) seep through the earth's crust. The resulting environment supports species not reliant on food and energy produced by photosynthesis.

selenizone A narrow, conspicuous, parallel-sided band of shell material that runs along the whorls of shells that have a slit. It originates at the back of the slit and has the same width.

septum A wall dividing a cavity or structure into smaller ones.

sessile Immobile. Attached to the substrate.

shell hash Substrate of coarsely broken shell.

shoulder An angulation in the curvature of the whorl, usually fairly close to the suture.

sigmoid S-shaped.

siphon A fleshy tube through which water enters the mantle cavity.

siphonal canal An elongated, semi-tubular extension of the aperture that protects the siphon of the snail.

siphonal notch A rounded notch at the anterior end of the aperture, through which the animal extends and retracts its siphon.

siphuncle A long, slender tube that passes through the chambers of cephalopod shells. Controls buoyancy.

sipunculan A phylum of bilaterally symmetrical unsegmented marine worms.

spicule A small spike or needle-like structure.

spire The portion of the shell between the apex and the body whorl.

stria (pl. striae) A shallow, incised groove in the surface of the shell.

stromboid notch A sinuous indentation along the outer lip of the shell, near the siphonal canal. Prevalent in the family Strombidae, but also present in other gastropods.

suctorial Adapted for sucking, either to draw fluid or adhere by suction.

sulcus A grove, furrow, or depression in the surface of a shell.

superfamily A group of related families, all of which are more closely related to each other than to any family outside the superfamily.

suprabranchial Above the gills.

supralittoral The zone above the high-tide line, which is regularly splashed by waves, but not normally submerged.

supratidal The area adjacent to and just above the high-tide line.

suspension feeder Animals that feed on material suspended in the water, usually by straining them from the water; also known as a filter feeder.

sutural laminae Anteriorly directed extensions of the articulamentum layer in Polyplacophora, extending beneath the posterior margin of the preceding valve.

suture A line along the shell surface along which adjacent whorls join.

symbiont An organism that is the beneficiary of a symbiotic relationship with another organism.

tegmentum The shell layer between the periostracum and articulamentum in Polyplacophora.

teleoconch That portion of the shell produced after the larva undergoes metamorphosis; the adult shell.

trema (pl. tremata) An opening in the shell through which water exits the mantle cavity.

trochophore The first larval stage in mollusks. Biconical in shape with two bands of cilia, before the shell is formed.

truncate Having a squared-off end.

tubercle A small, rounded, raised projection.

turbinate Shaped like an inverted cone.

turriculate In the form of a turret.

umbilicus A hollow, usually conical opening in the base of the shell. Present in shells in which the entire aperture, including its inner edge, is completely outside the coiling axis.

umbo (pl. umbones) The first part of a bivalve shell to form; the apex of each valve.

unguiculate Resembling a claw.

valve A distinct, calcified structure that forms all or part of the shell.

varix (pl. varices) A thickening along a lip of the shell, usually indicating an interruption in growth and reinforcement of the shell edge.

veliger A larval stage of mollusks, characterized by the presence of a velum, a membrane covered in fine cilia whose waving motions aid the larva's movement.

ventricose Distended, inflated, or swollen, especially on one side.

whorl A complete rotation (360 degrees) of the shell growth around the coiling axis.

wing The greatly extended ear on one side of the umbo of some bivalves.

FURTHER READING

Numerous books have been written about mollusks, from the days of Aristotle to the present. Some are intended for general readership, others for specialists and researchers.

The selected bibliography below, provides a small list of books on mollusks to introduce general readers to various aspects of shells and the mollusks than produce them. The listed websites will provide access to a far larger array of information at multiple levels of interest.

BOOKS ON MOLLUSKS

Abbott, R. Tucker. 1974.
American Seashells. Second edition.
Van Nostrand Reinhold Company: New York, NY.

Abbott, R. Tucker. 1989.
Compendium of Landshells.
American Malacologists, Inc.: Melbourne, FL.

Abbott, R. Tucker. 1993.
Kingdom of the Seashell.
American Malacologists

Abbott, R. Tucker & Dance, S. Peter. 1982.
Compendium of Seashells. Second edition.
American Malacologists: Melbourne, FL.

Barnett, Cynthia. 2021.
The Sound of the Sea: Seashells and the Fate of the Ocean.
WW. Norton & Co.

Beesley, Pamela L., Graham, J. B. Ross and Alice Wells. 1998.
Mollusca: The Southern Synthesis: Parts A & B: Fauna of Australia Volume 5.
CSIRO Publishing.

Bouchet, Phillipe & Mermet, Gilles. 2008.
Shells.
Abbeville Press.

Dance, S. Peter. 1969.
Rare Shells.
Faber & Faber, Ltd.

Dance, S. Peter. 1986.
A History of Shell Collecting.
E. J. Brill.

Dance, S. Peter, 2005.
Out of My Shell, A Diversion for Shell Lovers.
C-Shells-3, Inc.

Fearer Safer, Jane & McLaughlin Gill, Frances. 1982.
Spirals from the Sea: An Anthropological Look at Shells.
Clarkson N. Potter, Inc.

Harasewych, M. G. 1989.
Shells: Jewels from the Sea.
Rizzoli.

Harasewych, M. G. & Moretzsohn, Fabio. 2010.
The Book of Shells: A Life-Size Guide to Identifying and Classifying Six Hundred Seashells.
University of Chicago Press.

Landman, Neil H., Mikkelsen, Paula M., Bieler, Rudiger & Bronson, Bennet. 2001.
Pearls: A Natural History.
Harry N. Abrams, Inc.

Matsukuma, Akihiko, Okutani, Takashi & Habe, Tadashige. 1991.
World Seashells of Rarity and Beauty.
National Science Museum, Tokyo.

Meinhardt, Hans, 1995.
The Algorithmic Beauty of Sea Shells.
Heidelberg, Germany.

Mikkelsen, P. M., and R. Bieler. 2008.
Seashells of Southern Florida: Living Marine Mollusks of the Florida Keys and Adjacent Regions. Bivalves.
Princeton University Press, Princeton, New Jersey.

Moretzsohn, Fabio. 2023.
Shells: A Natural and Cultural History.
Reaktion Books.

Petuch, Edward J. 1994.
Atlas of Florida Fossil Shells.
Chicago Spectrum Press.

Ponder, W. F., D. R. Lindberg & J. M. Ponder. 2019.
Biology and Evolution of the Mollusca, Volume 1.
CRC Press, Boca Raton, Florida.
Volume 2. 2020. CRC Press, Boca Raton,
Florida.

Poppe, Guido T. 2008–2017.
Philippine Marine Mollusks, Volumes 1–5.
Conch Books.

Robin, Alain. 2008.
Encyclopedia of Marine Gastropods.
Conch Books.

Rosenberg, Gary. 1992.
The Encyclopedia of Seashells.
Dorset Press.

Stix, Hugh, Stix, Marguerite
& Abbott, R. Tucker. 1968.
The Shell, Five Hundred Million Years of Inspired Design.
Abrams, New York.

Sturm, C.F., Pearce, T. A. & Valdes, A. 2006.
*The Mollusks, A Guide to their Study,
collection, and preservation.*
American Malacological Soc.

Tenneson, Joyce. 2011.
Shells: Nature's Exquisite Creations.
Down East Books.

Vermeij, Geerat J. 1993.
A Natural History of Shells.
Princeton University Press.

Verrill, A. Hyatt. 1936.
Strange Sea Shells and their Stories.
L. C. Page & Co.

WEBSITES WITH INFORMATION ON MOLLUSKS AND SHELL COLLECTING

Conchologists of America (COA), a society for shell enthusiasts at all levels of interest
www.conchologistsofamerica.org/home

American Malacological Society, a society for researchers working with mollusks
www.malacological.org

Unitas Malacologia, an international society to further the study of mollusks
www.unitasmalacologica.org

National Shellfisheries Association, a society for studies on shellfish, including mollusks
www.shellfish.org

Mollia, sources of information for malacologists
www.ucmp.berkeley.edu/mologis/mollia

Broward Shell Club, one of several shell clubs featuring websites that provide links to information on mollusks for shell collectors.
browardshellclub.org

Jacksonville Shell Club, the website provides links and information on a wide variety of topics involving mollusks.
jaxshells.org

Seashell & Mollusc Links, a privately produced website with numerous links to national and international sites featuring various aspects of collecting and studying mollusks
www.petersseashells.com/shelllinks

Conchology, an extensive website with links to many topics involving mollusks, including price lists for shell books and specimen shells.
www.conchology.be

ConchBooks, the website of a publisher and dealer specializing in books on mollusks, both recently published and antiquarian.
www.conchbooks.de

INDEX

A

abalone 120–1
Acanthochitonidae 28–9
Acesta 59
Acharacinae 43
Acteonimorpha 200–1
Acteonoidea 200–1
Aculifera 11, 17
Adapedonta 76–9
adductor muscles 37–8, 41, 43, 46, 48, 50, 56–7, 58, 61, 63, 64, 66, 69, 70, 72, 76, 78, 80–1, 84, 91, 98
Admetinae 170–1
aesthetes 19, 24
ammonium chloride 219
Amygdalum 49
ancestral mollusk 11, 14
Anomalodesmata 70–3
Anomiidae 64–5
Anomioidea 64–5
Aplustridae 200–1
Aplysia californica 205
Aplysiida 204–5
Aplysiidae 204–5
Aplysioidea 204–5
apophysis 86–7, 130
Aporrhaidae 144–5
Aporrhais 144
Aporrhais pespelecani 145
aragonite 15
    bivalves 45, 48, 52, 61, 68, 72
    gastropods 165
    heteropods 165
    scaphopods 101
Architectonicidae 196–7
Architectonicoidea 196–7
Arcida 46–7
Arcidae 46–7
Arcinella 95
Arcoidea 46–7
Arctica islandica 7, 92
Arcticidae 92–3
Arcticoidea 92–3
Argonauta 222–3
Argonautidae 222–3
Argonautoidea 222–3
Aristotle 145
ark shells 46–7
Arrhoges 144
articulamentum 19
assassin snails 181
auger shells 192–3

Aulacoderida 220
Australiteuthidae 219
autotomy 58, 187
awning clams 42–3
axial ribs 37, 124, 133, 149, 160, 166, 168, 174, 178, 180, 182, 192, 198

B

bacteria
    chemoautotrophic 43, 96–7
    chemosynthetic 9, 48, 74–5, 96–7, 116–17
    symbiotic 9, 38, 43, 48, 74–5, 88–9, 96–7, 116–17
bait 84–5
baler shells 172–3
Bathymodiolus 48
Bathynerita naticoides 130–1
Bayerotrochus 118
Belemnitida 220
bioindicators 183
bioturbation 45
Bivalvia 17, 36–99
Blue Mussel 48
bonnet shells 158–9
Bouchetitrochus 118
broadcast spawning 14
brooch shells 66–7
Buccinidae 174–5
Buccinoidea 174–81
buoyancy 151, 211
    heteropods 164
    nautilus 214
    pencil squid 219
    Ram's Horn Squid 221
    siphuncle 212, 214, 216, 221
burrowing 45, 66, 73
    heart cockles 81
    olive snails 186, 187
    pelican's foot snails 145
    piddock clams 87
    razor shells 78–9
    shipworms 88
    surf clams 99
    Venus clams 91
Busyconidae 176–7
Busyconinae 176–7
Busycotypinae 176–7
byssal groove 49, 54, 61, 76
byssal threads 38, 54, 58, 60
byssus 46, 49, 56, 57, 61, 64

C

Caecidae 140–1

Caecum 141
caecum shells 140–1
Caenogastropoda 132–7, 166–7
calcite 48, 50, 52, 61
Callochitonida 24–5
Callochitonidae 24–5
Callochitons 24–5
Calyptraeidae 148–9
Calyptraeoidea 148–9
Cancellaria cooperii 171
Cancellariidae 170–1
Cancellarioidea 170–1
cap shells 150–1
captacula 103, 107
Capulidae 150–1
Capuloidea 150–1
Cardiida 80–5
Cardiidae 80–1
Cardioidea 80–1
Cardita clams 68–9
Cardita 68–9
Carditidae 68–9
Carditoidea 68–9
Carinariidae 164–5
carnivores 9
    auger shells 192–3
    bivalves 72
    cone shells 194
    dipper clams 72
    false cowries 154
    helmet shells 158–9
    heteropods 164
    microcarnivores 11, 106
    moon snails 156
    olive snails 186–7
    pleurotomariids 118–19
    pyram shells 199
    slit shells 118–19
    tritons 161
    true whelks 174–5
    tun shells 163
    turrid shells 191
    volutes 172
    wentletraps 166
carrier shells 146–7
Cassidae 158–9
Caudofoveata 11, 14
Cavoliniidae 206–7
Cavolinioidea 206–7
Cephalaspidea 202–3
cephalic eyes 53
cephalic shield 200, 202
Cephalopoda 17, 210–23
Cerithiidae 132–3

233

Cerithioidea 132–7
ceriths 132–3
Chama 94
Chamidae 94–5
Chamoidea 94–5
chank shells 184–5
chemoautotrophic bacteria 43, 96–7
chemosynthetic bacteria 9, 48, 74–5, 96–7, 116–17
Chitinoidea 26–7
Chitonida 26–33
Chitonidae 26–7
chitons 18–33
chromatophores 211, 216, 223
Chrysomallon 116, 117
Chrysomallon squamiferum 117
cigarette snail 194
clams 8, 9, 38
    awning 42–3
    Cardita 68–9
    dipper 72–3
    file 58–9
    giant 38, 80–1
    jewel box 94–5
    Lucina 74–5
    Ming the 92
    piddock 86–7
    surf 98–9
    Venus 90–1
    Yoldia 44–5
claspers 50
Clavagelloidea 70–1
climate change 207
Coleoidea 212, 218, 221
color
    ceriths 133
    chromatophores 211, 216, 223
    cowries 152
    Dentaliidae 105
    false cowries 154
    keyhole limpets 122
    nautilus 214
    sea hares 205
    surf clams 98
    tritons 161
    tulip shells 178
    volutes 173
    wedge shells 84
Columbariidae 184
columella 126, 130, 134, 146, 152, 158
    cap shells 151
    ceriths 132–3
    false cowries 154

lightning whelks 176
mud snails 180
paper bubble shells 200
pyram shells 198
top shells 124
true whelks 174
tulip shells 178
volutes 173
commensalism 57
commercial havesting 91
    abalone 121
    conches 143
    geoducks 76
    pearl oysters 55
    pencil squid 219
conches 142–3
Conchifera 11, 17, 34
Condylonucula maya 38
cone shells 194–5
Conidae 190, 194–5
Conoidea 190–5
conotoxins 194
Conus geographus 194
Coquina 85
cowries 7, 152–3
    false 154–5
Crassostreinae 50
Cremnoconchus conicus 139
Crepidula 148
Crucibulum 148
Cryptochitoninae 28
Cryptochiton stelleri 29
Cryptoplacidae 30–1
Cryptoplacoidea 28–31
Cryptoplax 30
Ctenoides 58
ctenolium 60
Ctenopelta 117
Ctiloceratinae 140–1
cultured pearls 54–5
cup and saucer shells 148–9
Cuspidariidae 72–3
Cuspidarioidea 72–3
cuttlefish 212, 216–17
Cycloneritida 130–1
Cymatiidae 160–1
Cypraeidae 152–3
Cypraeoidea 152–5

D

Decapodiformes 212, 218
defense 59
    foot autotomizing 187
    pleurotomariids 119
    sea hares 205

Dentaliida 103, 104–5
Dentaliidae 104–5
denticle 180
deposit feeding 45, 82, 134
detorsion 109
dipper clams 72–3
Donacidae 84–5
Donax variabilis 85
dyes 183

E

ecdysis 6
echinospira larvae 151
ectoparasites 154, 168, 198
egg cowries 154–5
eggs
    cephalopods 211–12
    paper nautilus 222
    Polyplacophora 20
    tritons 161
endoparasites 168–9
Enigmonia aenigmata 64–5
Entemnotrochus 118
epifaunal organisms 38, 62, 64, 94
epiproboscis 189
Epitoniidae 166–7
Epitonioidea 166–7
Epitonium scalare 166
Eudolium 163
Eulimidae 168–9
eulimid snails 168–9
Euthyneura 204–5
evolution 10–11
extremophiles 96
eyes
    cephalic 53
    pallial 53, 58, 63

F

false cowries 154–5
false limpets 208–9
Fasciolariidae 178–9
feeding 9
    deposit 45, 82, 134
    gastropods 110
    suctorial 170
    see also carnivores; filter feeders; herbivores; suspension feeding
file clams 58–9
filter feeders 9, 124
    brooch shells 66
    Cardita clams 68
    shipworms 88
    slipper shells 148, 149
    top shells 124

turret snails 134
Fissurellidae 122–3
Fissurelloidea 122–3
food for humans 48, 50, 84–5, 91, 143, 183
free living organisms 38, 95
funiculum 154

G

Gadilida 103, 104, 106–7
Gadilidae 106–7
Gastropoda 17, 108–69
geoducks 76–7
giant clams 38, 80–1
Gigantopelta 116
glassy bubble shells 202–3
Glossoidea 96–7
Green Mussel 48
gyroconic shell 221

H

Haliotidae 120–1
Haliotoidea 120–1
Haminoeidae 202–3
Haminoeoidea 202–3
hammer oysters 52–3
hat shells 148–9
heart cockles 80–1
hectocotylized arm 223
helmet shells 158–9
hemocyanin 41, 123
hemoglobin 43, 46, 49, 69, 75, 96
herbivores 9
  limpets 114
  turban shells 126
hermaphroditism 20, 50
  gastropods 111
  protandrous 50
  simultaneous 50
  turret snails 135
Heterobranchia 196–7
Heterochitoninae 32
heterodont hinge 74, 84, 91, 98
heteropods 164–5
heterostrophic protoconch 198
Hiatellidae 76–7
Hiatelloidea 76–7
hidden shell chitons 30–1
homing behavior 209
Horse Conch 178–9
Hydatinidae 200
hydrothermal vents 8, 9, 43, 48, 96, 174
  tapersnouts 116
  wood limpets 128

hypobranchial gland 119, 123, 166, 183, 189, 197
hypostracum 19

I

infaunal organisms 38, 45, 73, 77, 86, 90
infrapallial lobe 201
Inoceramidae 38
International Union for Conservation of Nature (IUCN) 9
introversion 106
iron sulfide 117
Isognomonidae 53

J

Janthina 166
Jason's Golden Fleece 57
jewel box clams 94–5
jingle shells 64–5

K

keyhole limpets 122–3
kleptoparasites 151
Krishna 184–5
Kuphus polythalamius 88–9

L

labial palps 41, 45, 73
Lambis 142
lappets 129
learning, cephalopods 211
Lepetellida 120–3, 128–9
Lepetellidae 122–3
Lepetelloidea 128–9
Lepidopleurida 22–3
Leptochiton 22
Leptochitonidae 22–3
lightning whelks 176–7
Lima 58
Limida 58–9
Limidae 58–9
Limnoperna 48
Limoidea 58–9
limpets 114–15
  false 208–9
  keyhole 122–3
  true 112–13
  wood 128–9
Lithophaginae 48–9
Littorinidae 138–9
Littorinimorpha 138–43, 144–65, 168–9
Littorinoidea 138–9

locomotion see swimming
Loliginidae 218–19
longevity 7, 47
  cuttlefish 216
  false limpets 209
  keyhole limpets 123
  nautilus 215
  ocean quahogs 92
  scaled chitons 26
  sea hares 204
  Venus clams 91
  whelks 175
Lophinae 50
Lottia gigantea 114
Lottiidae 114–15
Lottoidea 114–15
Lucina clams 74–5
Lucinida 74–5
Lucinidae 74–5
Lucinoidea 74–5

M

Mactridae 98–9
Mactroidea 98–9
Malea 163
Malleidae 52–3
Mantellina 59
mantle 7, 11, 15
  bivalves 38, 43, 45, 46, 50, 53, 54, 58, 61, 63, 64, 66, 69, 72, 74, 76, 78, 82, 92, 96
  cephalopods 211
  cowries 152
  false cowries 154
  keyhole limpets 122–3
  limpets 112
  Scaphopoda 101
mantle cavity 11, 14, 15, 38, 50, 53, 57, 63, 73, 78, 95, 101, 118
  false limpets 208
  glassy bubble shells 203
  keyhole limpets 122
  nautilus 215
  periwinkles 139
  top shells 124
  turban shells 126
  turret snails 134
mantle groove 19, 22, 24, 29
Margaritidae 55
mariculture 91
medicine 6, 152, 183, 194
Megathura crenulata 123
mentum 198
Mercenaria mercenaria 91
mesoplax 86

metamorphosis 8, 122, 140, 165
microcarnivores 11, 106
Mikadotrochus 118
Ming the Clam 92
miter shells 188–9
Mitridae 188–9
Mitroidea 188–9
molting 6
Monoplacophora 17, 34–5
moon snails 156–7
Mopaliidae 32–3
Mopalioidea 32–3
mother of pearl 54, 120–1
mucro 140
mud snails 180–1
murex shells 182–3
Muricidae 182–3
Muricoidea 182–3
mussels 48–9
Myida 86–9
Myopsida 218–19
Mytilida 48–9
Mytilidae 48–9
Mytiloidea 48–9
Mytilus edulis 48

N

nacre (mother of pearl) 54, 120–1
Nassariidae 180–1
nassa snails 180–1
Naticidae 156–7
Naticoidea 156–7
Nautilida 214–15
Nautilidae 214–15
Nautiloidea 212, 214–15
nautilus 214–15
Neogastropoda 170–95
Neomphalida 116–17
Neomphalioidea 116–17
Neopilinida 35
Neopilinidae 34–5
Neopilinoidea 35
nerites 130–1
Neritidae 130–1
Neritoidea 130–1
neurobiology model 205
neurotoxins 158, 159, 175, 194
nocturnal species 26, 153
Nodopelta 117
Nuculanida 44–5
Nuculanoidea 44–5
Nuculida 40–1
Nuculidae 40–1
Nuculoidea 40–1
nutmeg shells 170–1

nut shells 40–1

O

ocean quahogs 92–3
ocelli 46
Octopoda 222–3
Octopodiformes 212
octopods 212
octopuses 212
Oliva 187
olive snails 186–7
Olividae 186–7
Olivinae 186–7
Olivoidea 186–7
operculum 110
 auger shells 192
 caecum shells 140
 cap shells 151
 conches 143
 eulimid snails 168
 helmet shells 158
 lightning whelks 176
 limpets 112, 114
 moon snails 156
 mud snails 180
 nerites 130
 olive snails 186
 periwinkles 139
 pyram shells 198
 sponge worm shells 136, 137
 sundial shells 196
 tapersnouts 116, 117
 top shells 124
 tritons 161
 true whelks 174
 tulip shells 178
 tun shells 163
 turban shells 126
 turret snails 134
 turrid shells 191
opisthobranchs 109, 200
ornaments 6, 103, 143, 152, 159
osphradium 124, 143
Ostreida 50–5, 56–7
Ostreidae 50–1
Ostreinae 50
Ostreoidea 50–1
Ovulidae 154–5
oysters
 hammer 52–3
 pearl 54–5
 reefs 50
 thorny 62–3
 true 50–1
 wing 54–5

P

Pachydermia 116–17
pallets 88
pallial eyes 53, 58, 63
pallial line 38, 45, 48, 64, 69, 72, 74, 76, 78, 81, 82, 84, 95, 99
pallial sinus 38, 49, 72, 78, 82, 84, 91
pallial veil 53, 64
Panopea generosa 76–7
paper bubble shells 200–1
paper nautilus 222–3
papillae 45, 66, 73, 81, 121, 133
 awning clams 43
 cowries 152–3
 false cowries 154
 slit shells 118
 tooth shells 106
paralytic poisoning 187
parapodia 204, 206
parasites 9
 ectoparasites 154, 168, 198
 endoparasites 168–9
 eulimid snails 168–9
 kleptoparasites 151
Patellidae 112–13
Patellogastropoda 112–15
Patelloidea 112–13
pavillon 101, 104, 106
pearl oysters 54–5
Pecten jacobaeus 61
Pectinida 60–5
Pectinidae 60–1
Pectinoida 62–3
Pectinoidea 60–1
pedal gland 183
pelican's foot snails 144–5
Peltospiridae 116–17
pencil squid 218–19
Penicillidae 70–1
pen shells 56–7
periostracum 7, 19, 61, 66, 68, 72, 76, 82, 91, 98
 bivalves 37, 40, 42, 45, 46, 48, 54, 61, 66, 68, 72, 76, 80, 82, 91, 92, 96, 99
 cap shells 151
 carrier shells 146
 glassy bubble shells 202
 tapersnouts 116
 top shells 124
 tritons 161
 true whelks 174
 tulip shells 178
 tun shells 163

periwinkles 138–9
Perna viridis 48
pests 183, 199
Pharidae 78–9
Pholadidae 86–7
Pholadoidea 86–9
photophore 221
piddock clams 86–7
Pilgrim's Scallop 61
Pinctada 54–5
Pinctada imbricata 54–5
Pinctada margaritifera 54
Pinna nobilis 57
Pinnidae 56–7
Pinnoidea 56–7
Pleurotomariida 118–19
Pleurotomariidae 118–19
Pleurotomarioidea 118–19
Pliocardiinae 96
pneumostome 208
pollution 9
    bioindicators 183
Polyplacophora 11, 17, 18–33
Precious Wentletrap 166
predation see carnivores
prickly chitons 28–9
proboscis
    chank shells 184
    eulimid snails 169
    heteropods 164
    moon snails 156
    mud snails 180
    nutmeg shells 170
    pyram shells 199
    sundial shells 196
    true whelks 174
    tulip shells 179
    tun shells 163
    volutes 173
propodium 186
protandrous hermaphrodites 15, 50
protoconch 122, 128, 140, 161, 176, 196
    heteropods 165
    pyram shells 198
protoplax 86
Pseudochama 94–5
Pseudococculinidae 128–9
pseudoproboscis 151
Pteria 54
Pteriidae 54–5
Pterioidea 52–5
Pteriomorpha 58–9
Pteropoda 206–7

pteropod ooze 207
Pterotracheoidea 164–5
pulmonates 109
Pylopulmonata 198–9
Pyramidellidae 198–9
Pyramidelloidea 198–9
pyram shells 198–9

R

radula 19–20, 38, 101
    abalone 121
    auger shells 192–3
    cephalopods 211
    cone shells 194–5
    Dentaliidae 104
    keyhole limpets 122
    lightning whelks 177
    limpets 113, 114
    miter shells 189
    nutmeg shells 170
    paper bubble shells 201
    slipper shells 149
    sponge worm shells 136
    sundial shells 196
    top shells 124
    turban shells 126
    turrid shells 191
    volutes 173
Ram's Horn Squid 220–1
Rangia cuneata 98
razor shells 78–9
Recluzia 166
reproduction
    cephalopods 211–12
    gastropods 110–11
    see also eggs; hermaphroditism
rhinophores 204
rhynchodeum 192, 195
Rhynchopelta 117
rock shells 182–3

S

St James Scallop 61
scaled chitons 26–7
scallops 60–1, 151
Scaphopoda 17, 100–7
schizodont hinge 66
sea butterflies 206–7
sea hares 204–5
Sea Pangolin 117
selenizone 118
Sepiida 216–17
Sepiidae 216–17
Sepiina 216
Sepioidea 216–17

Sepiolina 216
sepion 216
septum 73, 130, 140
sexual dimorphism 177
    paper nautilus 222
    Ram's Horn Squid 221
shell-less classes 11, 14–15
shipworms 88–9
Sigapatella 148, 149
Siliquariidae 136–7
simultaneous hermaphrodites 50
Sinistrofulgur 176
Siphonariida 208–9
Siphonariidae 208–9
Siphonarioidea 208–9
siphons 38, 45, 48, 70, 72–3, 76, 78, 81, 82, 86–7, 88, 91, 96, 99
    dipper clams 72–3
    geoducks 76
    mud snails 180
    piddock clams 86–7
    surf clams 99
    Venus clams 91
    Vesicomyas 96
    volutes 173
siphuncle 212, 214, 216, 221
slender chitons 22–3
slipper shells 148–9
slit shells 118–19
Solemyida 42–3
Solemyidae 42–3
Solemyinae 43
Solemyoidea 42–3
Solenogastres 11, 15
Solenoidea 78–9
speckled chitons 32–3
spider conchs 7
spindle shells 178–9
Spirula 212
Spirula spirula 220–1
Spirulida 220–1
Spirulidae 220–1
Spiruloidea 220–1
Spisula solidissima 98
Spondylidae 62–3
Spondylus 63
sponge worm shells 136–7
squid 8, 211, 212
    pencil 218–19
    Ram's Horn Squid 220–1
Stellaria solaris 146
Strombidae 142–3
Stromboidea 142–3, 144–7
stromboid notch 142–3
suctorial feeding 170

sulfuric acid 158, 159, 163
sundial shells 196–7
surf clams 98–9
suspension feeding 40, 45, 53, 65, 77, 82, 95
    cap shells 151
    geoducks 77
    jewel box clams 95
    ocean quahogs 92
    sponge worm shells 136–7
swimming 8, 14, 38, 79
    file clams 58
    heart cockles 81
    heteropods 165
    pencil squid 218
    Ram's Horn Squid 221
    scallops 60
    sea butterflies 206
    sea hares 204
symbiotic algae 9
symbiotic bacteria 9, 38, 43, 48, 74–5, 88–9, 96–7, 116–17
Syrinx 184
Syrinx aruanus 184

T

tapersnouts 116–17
Tegillarca granosa 46
tegmentum 19
teleoconch 140
Tellinidae 82–3
Tellinoidea 82–5
tellins 82–3
tentacles
    cuttlefish 216
    Limidae 58
    limpets 112
    nautilus 214–15
    Scaphopoda 101
    sea butterflies 206–7
    wood limpets 129
Terebridae 190, 192–3
Teredinidae 88–9
territoriality 113
tetramine 175
thorny oysters 62–3
Tibia 142
Tonna 163
Tonnidae 162–3
Tonnoidea 158–63
tooth shells 106–7
top shells 124–5
torsion 109
toxins 187
    auger shells 192–3

cone shells 194–5
    neurotoxins 158, 159, 175, 194
    sea hares 205
    turrid shells 190
tremata 120–1
Tridacna 38
Trigoniida 66–7
Trigoniidae 66–7
Trigonioidea 66–7
Trisidos 46–7
tritons 160–1
Trochida 124–7
Trochidae 124–5
Trochita 148, 149
Trochoidea 124–7
true limpets 112–13
true oysters 50–1
true scallops 60–1
true whelks 174–5
trumpets 185
Truncatelloidea 140–1
Tudivasum 185
tulip shells 178–9
tun shells 162–3
turban shells 126–7
Turbinella 184
Turbinella pyrum 184
Turbinellidae 184–5
Turbinellinae 184
Turbinelloidea 184–5
Turbinidae 126–7
turret snails 134–5
Turridae 190–1
turrid shells 190–1
Turritellidae 134–5
tusk shells 101, 104–5
Tyrian purple 183

U

umbilicus 146, 156
umbone 52, 64, 74, 76, 78, 91, 98
Umbonium 124

V

Vampire Snail 171
Vanikoroidea 168–9
vase shells 184–5
Vasinae 184, 185
Vasum 185
Venerida 90–9
Veneridae 90–1
Veneroidea 90–1
Venus clams 90–1
Vesicomya 97
Vesicomyas 96–7

Vesicomyidae 96–7
Vesicomyinae 97
Vetigastropoda 118–29
volutes 172–3
Volutidae 172–3
Volutoidea 172–3

W

wampum 91
watering pot shells 70–1
wedge shells 84–5
wentletraps 166–7
whelks
    lightning 176–7
    true 174–5
wing oysters 54–5
wood limpets 128–9

X

Xenophoridae 146–7

Y

Yoldia clams 44–5
Yoldiidae 44–5

PICTURE CREDITS

The publisher would like to thank the following for permission to reproduce copyright material:

Alamy/Agefotostock: 24, 83 (top), 91 (top); Auscape: 189 (top); Biosphoto: 105, 146; Blickwinkel: 89 (top), 149; Blue Planet Archive: 223 (bottom); Laurie Campbell: 139 (left); Dorling Kindersley: 82, 135 (left), 174, 185 (bottom); Jason Edwards: 27 (bottom); David Fleetham: 203 (top); Frank Hecker: 175 (right); ImageBROKER: 69; Klepo: 56; MShieldsPhotos: 177 (top); Nature Photographers Ltd: 87 (left), 151 (right); Andrey Nekrasov: 138, 181 (top); Roberto Nistri: 208; Reef and Aquarium Photography: 133; Lee Rentz: 65 (top); Stocktrek Images, Inc.: 168; Marli Wakeling: 126; Adrian Weston: 112; Doug Wilson: 77. Ardea/Paulo Di Oliviera: 195 (top). Bailey-Matthews National Shell Museum/Dr. José H. Leal: 40, 50, 65 (bottom), 71, 74, 95, 98, 99 (centre), 170, 203 (centre), 203 (bottom). SN Barid, CC BY-SA: 134. Zachary Berghorst: 140. Vishal Jayant Bhave: 185 (top). Philippe Bourjon, CC BY-SA: 75 (top). © Marilynne Box, CC BY: 23. www.britishmarinelifepictures.co.uk/Jason Gregory: 93 (bottom). © 2014 Checa et al. Checa AG, Salas C, Harper EM, Bueno-Pérez JdD (2014) Early Stage Biomineralization in the Periostracum of the 'Living Fossil' Bivalve Neotrigonia. PLoS ONE 9(2): e90033. https://doi.org/10.1371/journal.pone.0090033. CC BY: 67. Chong CHEN: 117. Roger Clark: 33 (bottom). Gary Cobb: 201. Dreamstime/Xiao Zhou: 161 (top), 161 (bottom). Jonas Drotner Mouritsen: 218. Frédéric Ducarme, CC BY-SA: 63. Anne K DuPont: 205 (top left). © Femorale: 22, 25 (bottom), 163, 184. Florida Museum of Natural History/Gustav Paulay: 14 (bottom), 45. Bill Frank; with permission from jaxshells.org, a subsidiary of Conchologists of America, Inc.: 81 (bottom), 177 (bottom). Anne Frijsinger & Mat Vestjens – Natuurlijkmooi.net: 41. Getty Images/Auscape: 94; De Agostini: 53; DEA/Bartella: 144; Stephen Frink: 60; © Francesco Pacienza: 108; Paul Starosta: 2; 16 (top left); 16 (bottom left); 16 (top right); 16 (bottom right); 85 (centre); 171 (top). Gonzalo Giribet: 43. Hakai Institute: 73. M. G. Harasewych: 7, 26, 29 (top), 32, 44, 46, 57 (top), 57 (bottom), 59 (top), 59 (bottom), 61 (top), 61 (bottom), 62, 66, 72, 78, 80, 88, 90, 92, 96, 99 (bottom), 110, 114, 118, 121 (top left), 123 (top), 123 (bottom), 125 (bottom left), 125 (bottom right), 128, 131 (left), 135 (right), 143 (top left), 143 (bottom), 173 (bottom left), 173 (right), 175 (left), 179 (bottom left), 179 (right), 182, 183, 193 (top right), 195 (bottom right), 196, 220, 223 (top). iStock/JustineG: 85 (top). © Javier, CC BY: 25 (top). Jeanette & Scott Johnson: 182, 190, 192. © Paul Kay: 93 (top). Scanning electron micrographs by Kevin Kocot: 14 (top), 15 (top). Lima SFB, Christoffersen ML (2016) Redescription and designation of a neotype for *Caecum floridanum* (Littorinimorpha, Truncatelloidea, Caecidae) with a characterization of the protoconch and growth stages. ZooKeys 585: 17-31. https://doi.org/10.3897/zookeys.585.7646. CC BY: 141 (top right). © Saryu Mae, CC BY: 137 (top). MARUM – Center for Marine Environmental Sciences, University of Bremen: 49 (top). © Erin McKittrick, CC BY: 150. © MNHN, Paris: 15 (bottom), 30, 152; Manuel Caballer, CC BY: 68, 116, 129 (top), 207 (top); Velásquez and Shipway 2018, CC BY: 89 (bottom). Museos Científicos Coruñeses, CC BY-SA: 54. Muséum de Toulouse/Didier Descouens, CC BY-SA: 55 (bottom). Museum of New Zealand Te Papa Tongarewa, CC BY: 35 (top), 42, 129 (bottom). Nature Picture Library/Ingo Arndt: 127 (top left), 147 (top), 147 (bottom); Ingo Arndt/Minden: 159 (left), 167 (bottom); 173 (top left); Franco Banfi: 160; Fred Bavendam/Minden: 215; Philippe Clement: 79; Christophe Courteau: 28, 113 (bottom), 180; Georgette Douwma: 51 (bottom), 58, 166, 217; Guy Edwardes: 11; David Fleetham: 153 (top); Juergen Freund: 155 (top), 225; Shane Gross: 12 (bottom), 142; David Hall: 18, 33 (top), 157 (top); Sergio Hanquet: 130; Jaymi Heimbuch: 29 (bottom); Magnus Lundgren: 222;

Magnus Lundgren/Wild Wonders of China: 205 (bottom); Alex Mustard: 219; Pete Oxford/Minden: 81 (top); D. Parer & E. Parer-Cook/Minden: 8; Doug Perrine: 158; Constantinos Petrinos: 157 (bottom); Michael Pitts: 178; Jeff Rotman: 6; Jose B. Ruiz: 27 (top); Peter Scoones: 210; David Shale: 164, 206; Wild Wonders of Europe/Lundgren: 125 (top left); Norbert Wu/Minden: 48; Tony Wu: 115 (top). NOAA/Brooke et al, HBOI: 119 (top); Claire Fackler, CINMS, CC BY: 122. Sue Peatling: 31 (top). © Guido & Philippe Poppe – www.conchology.be: 9 (top), 55 (top), 75 (bottom), 85 (bottom), 86, 91 (bottom), 103 (bottom), 103 (top), 104 (left), 107 (top right), 107 (bottom right), 132 (right), 141 (top left), 155 (bottom right), 165, 189 (bottom), 195 (bottom left), 200, 207 (bottom), 209, 216. © Philippe & Guido Poppe – www.poppe-images.com: 15 (bottom), 120, 152, 153 (centre), 162, 171 (bottom), 172, 188, 197. Denis Riek: 169 (right). Harry Rose, CC BY: 47. Imagery courtesy of the Schmidt Ocean Institute: 221. Science Photo Library/British Antarctic Survey: 100; DK Images: 10; Harry Rogers: 84; Alexander Semenov: 36; Smithsonian Institution: 20, 156; Wim Van Egmond: 9 (bottom). © Ian Shaw: 167. Shutterstock/Andersphoto: 214; Cavan Images: 51 (top); Gerald Robert Fischer: 204; David A Litman: 12 (top); Leo Lorenzo: 154. Smithsonian Institution, Department of Invertebrate Zoology: 76, 83 (bottom), 107 (left), 119 (bottom), 127 (bottom right), 169 (left), 193 (bottom right), 194. SuperStock/Universal Images: 31 (bottom). © Christina Sylvester: 198. Richard Toller, CC BY-ND: 157 (centre). Amy Tripp: 87 (right), 179 (top), 186. Wiklund H, Taylor JD, Dahlgren TG, Todt C, Ikebe C, Rabone M, Glover AG (2017) Abyssal fauna of the UK-1 polymetallic nodule exploration area, Clarion-Clipperton Zone, central Pacific Ocean: Mollusca. ZooKeys 707: 1-46. https://doi.org/10.3897/zookeys.707.13042. CC BY: 35 (bottom), 106. www.wildsingapore.com/Ria Tan: 52, 64, 70, 99 (top), 124. © Woods Hole Oceanographic Institution/DSV Alvin: 97. Joel Wooster; with permission from jaxshells.org, a subsidiary of Conchologists of America, Inc.: 205 (top right). H. Zell, CC BY-SA: 5, 49 (bottom), 104 (centre), 104 (right), 113 (top left), 113 (top centre), 113 (top right), 115 (bottom), 121 (top right), 121 (bottom), 125 (top right), 125 (centre right), 127 (left), 131 (right), 132 (left), 136, 137 (bottom), 139 (top centre), 139 (top right), 139 (bottom), 141 (bottom), 143 (top right), 145, 148 (left), 148 (right), 151 (left), 153 (bottom), 155 (bottom left), 155 (bottom centre), 159 (top), 159 (bottom), 176, 181 (bottom left), 181 (bottom right), 187, 191 (top), 191 (bottom), 193 (top left), 193 (top centre), 193 (bottom left), 199 (left), 199 (right), 202.

The publisher gratefully acknowledges the following publications as sources for illustration references: Beesley, Pamela L., Graham J. B. Ross and Alice Wells. 1998. *Mollusca: The Southern Synthesis: Parts A & B: Fauna of Australia* Volume 5. CSIRO Publishing. Mikkelsen, P. M., and R. Bieler. 2008. *Seashells of Southern Florida: Living Marine Mollusks of the Florida Keys and Adjacent Regions. Bivalves*. Princeton University Press, Princeton, New Jersey. Ponder, W. F., D. R. Lindberg and J. M. Ponder. 2019. *Biology and Evolution of the Mollusca*, Volume 1. CRC Press, Boca Raton, Florida. Volume 2. 2020. CRC Press, Boca Raton, Florida.

All reasonable efforts have been made to trace copyright holders and to obtain their permission for the use of copyright material. The publisher apologizes for any errors or omissions and will gratefully incorporate any corrections in future reprints if notified.

ACKNOWLEDGMENTS

I am grateful to my many colleagues throughout the world and to the numerous enthusiastic shell collectors and dealers who generously shared information and provided images of living mollusks as well as shells, many of which appear in this book. Guido and Philippe Poppe of Conchology, Inc., Marcus and José Coltro of Femorale Shells, Dr. José Leal of the Bailey Matthews National Shell Museum, and Dr. Harry Lee of the Jacksonville Shell Club have been especially helpful in making images on their websites available.

Special thanks to Joanna Bentley, Ruth Patrick, Jacqui Sayers, Caroline Elliker, Katie Greenwood, and Kevin Knight of Bright Press for their many contributions to the content, design, and organization of *Shells of the World: A Natural History*.

I especially thank my wife and daughters for their support and patience during the production of this book.

ABOUT THE AUTHOR

M. G. Harasewych is a research zoologist and was a Curator of Mollusca at the National Museum of Natural History, Smithsonian Institution, in Washington, D.C. which houses the largest collection of mollusks in the world, from 1985 until his retirement in 2015. He remains an Emeritus Curator at this museum and continues to research the evolutionary relationships and ecology of various gastropod lineages, some inhabiting tropical islands, others abyssal plains off Antarctica. He has conducted field work in many regions of the world and often used research submarines to observe and sample mollusks from the ocean depths.

Over the years he has published more than 150 research papers and described dozens of species and genera of living and fossil mollusks. He authored *Shells: Jewels from the Sea* (1989) and co-authored with Fabio Moritzsohn *The Book of Shells* (2010) and *The Little Book of Shells* (2020) for general audiences. He served as Editor of *The Nautilus*, the oldest malacological journal in the western hemisphere, was President of the Biological Society of Washington, and is a Fellow of the American Association for the Advancement of Science and of the Explorer's Club.